se for

I live in the future & here's how it works

"Nick Bilton has written a rollicking, upbeat guide to the digital world—a peek into our near future, where news, storytelling, and even human identity are transformed. It's a fascinating book from a man who has helped pilot the *New York Times* into a new age of online journalism. If you're wondering—or worried—about the future of media, this is your road map."

> —**Clive Thompson,** *Wired* **magazine columnist and contributing editor**

"A bold and provocative look at the future of storytelling. It's about the virtues of videogames, the science of cocktail parties, and the new business model of journalism. It's about a world in which the medium is mostly irrelevant, and the message is everything. Read this book if you want to get your message right."

> —**Jonah Lehrer, author of the** *New York Times* **bestseller** *How We Decide*

"Bilton doesn't just live in the future, he also understands the past. *I Live in the Future* explains how our communications tools shaped our present, how new tools are shaping our future, and what we should do to take advantage of this opportunity."

> —**Clay Shirky, author of** *Cognitive Surplus* **and** *Here Comes Everybody*

i
live in the future &
here's how it
works

why your world, work, and brain are being
creatively disrupted

Nick Bilton

CROWN
BUSINESS
NEW YORK

Copyright © 2010 by Nick Bilton

Published in the United States by Crown Business, an imprint of
the Crown Publishing Group, a division of Random House, Inc., New York.
www.crownpublishing.com

CROWN BUSINESS is a trademark and CROWN and the Rising Sun
colophon are registered trademarks of Random House, Inc.

Originally published as a hardcover in the United States by Crown Business,
an imprint of the Crown Publishing Group, a division of Random House, Inc.,
New York, in 2010.

Crown Business books are available at special discounts for bulk purchases for
sales promotions or corporate use. Special editions, including personalized covers,
excerpts of existing books, or books with corporate logos, can be created in large
quantities for special needs. For more information, contact Premium Sales at
(212) 572-2232 or e-mail specialmarkets@randomhouse.com.

Library of Congress Cataloging-in-Publication Data
Bilton, Nick.
I live in the future and here's how it works / Nick Bilton.—1st ed.
p. cm.
Includes index.
1. Technological forecasting. 2. Technology—Social aspects. 3. Computers
and civilization. 4. Ubiquitous computing. I. Title.
T174.B53 2010
303.48'34—dc22   2010026870

ISBN 978-0-307-59112-8
eISBN 978-0-307-59113-5

Printed in the United States of America

Design by Lauren Dong
Cover photograph © Image Source/SuperStock

1 3 5 7 9 10 8 6 4 2

First Paperback Edition

for danielle

i <3 u

contents

author's note

Dear Reader,

This is not just a book but a unique reading experience.

Online, through a computer or smart phone, you can access
additional content for each chapter: videos, links to articles and
research, and interactive experiences that enable you to delve
deeper into the topics covered in that chapter, taking you beyond
the printed page.

At the beginning of each chapter you will see an image called
a QR Code, just like the one above. Using a free application you
can download from nickbilton.com you will be able to snap an image

of these codes that will then take you to the additional content directly on your mobile phone.

Become part of the *I Live in the Future* community by commenting on chapters of interest and joining a continuing discussion with me and your fellow readers online at nickbilton.com and with the free *I Live in the Future* app for iPhone and iPad.

introduction
cancel my subscription

As you will see, I eat my own dog food.

I used to love reading print newspapers. In 2004, when I started working at the *New York Times,* I was excited beyond words to discover that much of the Sunday *Times* was printed ahead of time and a stack of those early-run papers arrived at the *Times* building every Saturday. Not only did I work at one of the most respected newspapers in the world, but along with a paycheck, I also got the magazine, the Week in Review, the Metro section, and Sunday Business several hours before the rest of the world!

A new favorite ritual took root: I'd head to the office early every Saturday afternoon, and when the first delivery trucks arrived, I'd grab a few smudged copies and run home to immerse myself in tomorrow's newspaper. Before long, friends began calling me to ask for advance copies of the real estate section or the Sunday magazine.

Then, a couple of years later, I stopped my Saturday routine. The calls stopped too. One by one, my friends were switching to new reading rituals, replacing the smell and feel of the printed page with a quicker, personally edited, digital reading experience. Even when the paper was free, they didn't want a copy anymore!

The same thing was happening to me. I had started reading newspapers in high school and for years had stumbled every morning to the doorstep, blurry-eyed and half asleep, to fetch the morning paper. But now I was checking the headlines in the morning on my computer, reading articles on my mobile phone on the way to the office, and surfing news sites all day long. Aided by social networks such as Facebook and Twitter that helped pull together the best content at a vastly quicker pace, I now could see news more quickly online. I also had a much easier and more succinct way to share the articles I found interesting while adding my own commentary, helping to cull the best morsels of content for my friends, family, and coworkers. In retrospect, I was going through a personal "digital metamorphosis"—something many of you will experience, if you haven't already. For some, it will happen over time as you move one paper task after another to the computer, phone, or digital reader. For others, it will happen quickly with the purchase of a fancy new phone or new reader that suddenly opens up a whole new world of electronic possibilities.

In my case, unread newspapers at home began to climb to furniture-sized proportions by the front door, with the bottom layer turning a sickening shade of khaki yellow. My wife and I simply referred to the growing tower as the Pile.

Eventually, as the yellowing newspapers continued to collect, I decided it was time to take the plunge. I waited until lunchtime to make the call, checking the sea of cubicles around me to make sure nobody could hear me. I felt like a philandering spouse, and the idea of being a cheater didn't feel good.

I picked up the phone and called the *Times* circulation department. I even tried to disguise my voice in case someone recognized me, adding a tinge of an accent and speaking a little more slowly.

"Yes, I'm sure I want to cancel the delivery," I told the rep. "I'm sorry, I just don't read it anymore."

Of course, I love the *New York Times*. The stories are still top notch, as good as they've ever been: perceptive, exploratory, thoughtful, and informative. The problem is that the approach just doesn't make sense to me anymore. I understand the concept—the printed paper is a neat package with a hundred or so news articles, displayed by subject and order of importance, culled by *Times* editors, my colleagues. Top stories are here, business articles are there, sports is in the back of the business section on most weekdays.

But that's the problem: It's only a collection of what editors think is appropriate. And it doesn't swirl in my preferences. My likes and dislikes; it's just not designed for me. More important, by the time those carefully chosen words on paper arrive at my house, printed permanently on the page and selected for a vast audience of readers, a lot of the content isn't current.

A few years passed while I contentedly consumed the news in my own way. I continued to do my work at the New York

Times Research Labs, helping the Old Gray Lady find her place in mobile phones, on the computer screen, and in video, and my workplace infidelity remained my own private business. Then, in spring 2009, I appeared on a roster of speakers for the geeky O'Reilly Emerging Technology Conference in San Jose, California, aimed at cutting-edge technology developers. A *Wired* magazine reporter attending the conference asked for an interview.

Like a good corporate citizen, I checked with the *Times* public relations folks to make sure the interview was OK. Once they gave the go-ahead, I sat down with reporter Ryan Singel.

For over an hour, I showed Singel some of the prototypes from the *Times* research labs, such as the inner workings of our digital living room, where content can move seamlessly from my computer to a phone and back to a big-screen television. I showed him how videos on my computer of cookbook author and "Minimalist" columnist Mark Bittman whipping up a dish can appear instantly on my television while the recipe pops up on my phone. Every device could be connected to the others, and the stories I read on the computer could be illustrated with maps or video interviews on the TV, computer, or phone. Some day, I explained, sensors in the couch might alert the television or the computer to turn to my favorite shows or sites, or sensors in my phone might detect when I'm in the car and prompt information to be read aloud instead of displayed. For those who still want to read on paper, newspaper boxes might print out a personalized version—with customized advertising and even the ability to notify a nearby Starbucks that I was headed in for coffee.

I talked excitedly about some of our prototype mobile

applications in which the news could change on the basis of various scenarios. Imagine walking down a city block at lunchtime while reading the *Times* on a smart phone; since the phone knows it's lunchtime, articles related to food and local restaurants could appear. I showed him prototypes and concepts of flexible displays in which a bendable screen is constantly updating the news and can be folded away like a piece of paper.

At the very end of the interview, as Singel was getting ready to leave, he asked if I read the print paper. I was briefly unsure how to answer. Should I lie? The decision had been made so long ago that I hadn't recently considered the consequences of canceling my subscription. But it was now 2009, the age of netbooks, iPhones, and Kindles. I decided to be honest: I told him I mostly enjoyed reading the *New York Times* on my computer, mobile phone, and e-reader.

A few hours later I gave my presentation, chatted with a few interested attendees, and went back to my hotel room to discover my e-mail inbox crammed with messages. Some friends and coworkers in the newsroom were congratulatory. "Hey, Nick, great article on wired.com!" they wrote. "It's really great to see the NYTimes get so much digital credit."

But others, from coworkers on the business side of the company, had an ominous tone: "Holy shit, people here are pissed!"

"The grown-ups are talking," one said simply.

I was mystified about what I possibly could have said to get the grown-ups talking, so I went to wired.com. Under the headline "*Times* Techie Envisions the Future of News," with a nerdy picture of me smiling with my laptop, ran this:

"Nick Bilton, an editor in the *New York Times* research

and development lab, doesn't think much of newspaper[s]. In fact, he doesn't even get the Sunday paper delivered to his house.

"Thankfully for Bilton and his employer, he's bullish on news." Continuing, Singel added, referring to my feeling about paper, not about the *Times*, "It's just the paper he hates."

After this opener, Singel gave a concise and overwhelmingly positive overview of the work I showed him from our lab. The article was supportive of our work and should have been great coverage for a company aiming to show its shareholders that it is truly a forward-thinking digital organization. Some of my colleagues were thrilled that the story demonstrated how the paper was focusing on the future.

But some of my coworkers and bosses were incensed that I had publicly confessed to shunning the core product of the *Times*. Some even believed that I might persuade other readers to cancel their subscriptions as well.

When I returned to the New York office the next day, I was immediately informed that I shouldn't be telling the world that I don't read the print version. To quell some of the trauma, I apologized for my remarks.

In all honesty, however, I was completely confused. Clearly, I wasn't the only person who had stopped reading the print edition. In fact, what has happened nationwide in the last few years is truly shocking: In 2008, paid newspaper circulation in the United States fell to 49.1 million, the lowest number since the late 1960s and well below the peak of 60 million reached in the 1990s, when the Internet was just starting to come into its own. The *Times* has suffered as well, with circulation sliding in the 1990s, leveling off in the early part of the century, and then sliding some more. Daily circulation, which had been close to 1.2 million in the early 1990s, was close to 1 million at the time of my speech and would slip below the seven-figure mark later in 2009.

Print circulation told only part of the story. With a deep and painful recession accompanying a technological shift, advertisers have abandoned print papers even faster than subscribers have. Industrywide, revenue from print advertising has fallen off a cliff, plunging to $24.8 billion in 2009 from $47.4 billion in 2005, according to the Newspaper Association of America. That's a decline of nearly half in five years.

Newspapers are far from the only medium to face such agonizing declines. The digital revolution is roiling just about every form of media we know: Book sales in 2009 slipped to the lowest level since 2004, according to the Association of American Publishers. The Publishers Information Bureau reported that although magazine subscriptions have grown slightly, advertising pages sold dropped more than 25 percent in 2009. Despite the growing popularity of Blu-ray discs and a healthy box office, DVD sales fell 8 percent in 2008. The music industry has been hit hardest of all. Worldwide dollar sales have fallen every year for a decade—and the bottom is nowhere to be found. In 2009, CD sales fell more than 20 percent in both dollars and units. Although digital downloads are up and now account for about 40 percent of music sold, the revenue they bring in doesn't begin to make up for the disappearing disk sales.

Given this revolutionary shift in how we read, listen, and enjoy entertainment, shouldn't the *Times* be asking why I prefer digital to print and exploring how I consume my news? Shouldn't we be moving forward and not backward?

Imagine that you owned a restaurant and offered your employees free food, but they instead brought their own lunch and dinner from home. Would you look the other way if plates of freshly cooked pasta and garlic bread sat untouched on the table? Hopefully not. If it were my restaurant, I'd want to know why they weren't enjoying my product, and I would do everything I could to try to change that.

At Google they call this "dogfooding." That is, if you make dog food and the dogs won't eat it, you might have a bit of a problem. The people who built Gmail have to use it for their e-mail service, and if something doesn't work, they have to fix it. Collectively, if Google engineers don't like a service's feature, they are supposed to change it accordingly—whether it's Google Search, Google Mobile, or any other Google product. Along the same lines, if I wasn't reading the print newspaper, there was a reason.

Still, my published comments didn't end with that slap on the wrist. I heard from numerous people from numerous departments numerous times. But at each turn I continued to push at the issue. The conversation shouldn't be about my remarks in public, I insisted, but about my actions. I wanted to point out that with regard to the new delivery methods and the next generation's consumer habits, the writing was on the wall—or the screen, if you will.

I tried to explain that I—like many in my generation—preferred the instantaneous digital experience because I could share my favorite articles with others, adding comments and joining a collective discussion while also viewing other readers' opinions. The print paper is static, and so is its narrative; in comparison, a digital narrative can include invigorating interactive multimedia such as videos and slide shows. I also explained that people in my social networks and those I trusted shared relevant content with me, and their remarks and news gathering had become a critical filter for the stories I consumed. It wasn't about print versus digital; it was about immediacy, details, links, interactive graphics, videos, and, most important, hyperpersonalization. The majority of news I consumed was still from the *Times*. I just consumed it in a different way.

Although I didn't want to be insolent, they needed to accept that and respond to it. My peers aren't going to wake up one day and crave newsprint. The world is shifting; ignoring it won't make it go away.

The whole experience was the least enjoyable—and most anxious—of my six years at the *Times*. Thankfully, most of the pressure subsided after a few weeks—although I'm pretty sure there were some corporate suits who would have been happy to see my exodus from the company with a box of my belongings in hand. Luckily for me, and for the *Times*, this group is in the minority, and the paper of record continues to push at the forefront of the digital reshaping of news, aptly illustrated by the fact that I worked in a research lab and am visible to the public by the extraordinary journalism, innovation, and cutting-edge digital content the *Times* puts out on a daily basis.

I should add here that if you still read the news on paper, that's perfectly OK. Paper is still gadget number one for reading content; it's disposable, relatively inexpensive, and relatively simple to create in small or large quantities, and it doesn't need batteries or a power outlet. Admittedly, the online experience still isn't better than that of paper, and it has a long way to go until it is.

But paper alternatives are coming, and in some situations they are already here. Technology companies are working to make every aspect of our lives sync up with the digital world. Global positioning systems are replacing maps, grocery coupons appear on your phone, and the online phone directory is far more efficient than your local phone book. Eventually, a paper replacement for your daily news will come along too. This book will help you understand what this all means and how you can respond.

I Live in the Future

Granted, I'm a geek. I grew up playing the first video games ever made, and I still get excited by anything with buttons or a screen. I'm also hardwired for this wireless world. Call it ADD, impatience, or an overactive imagination, but I've always found it very tough to concentrate on just one topic.

My career path reflects this. I started out in the movie industry designing film titles. Then I moved to packaging design, where I created the initial mock-up for the first ever Britney Spears doll. (Please don't hold that against me—we all do things we're not proud of!) From packaging, I moved into advertising, which quickly morphed into Web advertising and Web programming. When the dot-com bubble burst in 2000, I decided to become a documentary filmmaker. I entered a yearlong certificate program in journalism and documentary film at New York University and then switched careers again, working at smaller alternative weekly newspapers in New York, where I learned the ropes.

My first job at the *Times* was as the art director of the Business and Circuits sections. Soon enough, my boss found out that I could both write stories and write computer code, and I was secretly assigned to a new digital reading collaboration project between Microsoft and the *Times*. (The project, called Times Reader, built a new kind of digital newspaper for tablet computers.) From there, I moved into two new research and technology-integration roles. For three years, I was the user interface specialist and researcher in the research and development department at the New York Times Company. The R&D Labs, as they are called, focused on a variety of projects, including building and prototyping mobile phone applications and working with device manufacturers to try to influence the boundaries of e-readers and the coming flexible screens. We also wrote short "white papers" for the company, exploring and explaining the implications of unlimited wireless Internet or doing informed speculative research on upcoming technologies and how they will affect the way we create, consume, and deliver content in the next few years. Our core mission in R&D was looking into the future to try to forecast how the technology and media worlds will work in the next two to ten years— what gadgets we'll use, the media we will consume, and what advertising will accompany those channels.

Simultaneously, I worked in the newsroom as design integration editor, charged with rethinking how the print narrative can morph and adapt to a digital form. More recently, I've joined the business section writing staff as the lead blogger for Bits, the paper's technology blog.

When I looked at all the different jobs I've been involved with over the last fifteen years—from advertising, writing, and photography to video, programming, and user interface design—I noticed one undeviating thread that ties it all together: storytelling.

All the pieces of my work—the photos, the words, the packages, the design, the programming code—all work hand-in-glove to tell a story. In fact, many of you are storytellers too, using a variety of media and marketing to sell your products, your political candidates, or simply your best ideas. Everything we do, in one form or another, is storytelling.

Just like me, the generation coming of age in this digital society doesn't see or perceive much difference in types of media. Video? Words? Music? Computer code? It doesn't matter. The actual tools used are irrelevant. It's the end result—the storylines, the messages—that matters. This generation thinks in pictures, words, and still and moving images and is comfortable mixing them all in the same space.

Even more, they don't need professionals or professional equipment to make it happen or direct it. With a computer and an inexpensive camera, they can create and consume in short, medium, and long forms. And if a form doesn't exist, they can create it. They are the new regime of storytellers.

You, Too, Will Be in the Future Soon Enough

It wasn't so long ago that content of all kinds seemed to be packed into big heavy bundles. You didn't buy a great story; you bought a magazine or a book. For the most part, you bought albums, cassettes, or CDs, not single songs. Movies were an evening's entertainment. The only editing was done by professionals, and distribution was handled by large companies with skilled salespeople and deep marketing budgets. Everything was sold at a markup, though in some cases advertising subsidized the cost.

Not anymore. Today, driven by a surge in technological innovation, that model is caving in on all sides. Look at computers as an example: As memory, storage capacity, and screens have become less expensive, the options have grown beyond the wildest dreams of a quarter century ago. The byte—the computer's single unit of data—was grouped in mere thousands in the 1980s to create games so basic that they were simply dots, lines, and equations. Today, video games are so real that it's hard to tell if you're watching a movie or playing in a virtual world.

The pricing of these technologies also tells a fascinating story: In 1984, the 10-megabyte hard drive was a wide-eyed wonder and considered a real deal at $4,495. By 2004, just twenty years later, such a drive was completely obsolete, too small to be used for modern computing tasks and not worth the effort to make. Today, $100 will easily buy you more than 500 gigabytes of storage—50,000 times as much storage space for a fraction of the price.

These kinds of stunning advances are driving many of the changes that are upending just about every form of media we know. Gradually, as the costs decrease, smart screens will begin to replace everything else, becoming all-purpose displays for TV shows, newspapers, blogs, Facebook status updates, family photos, magazines, and books. Content companies won't be confined to any one purpose, and they will be able to create and distribute virtually any kind of information or entertainment in all sizes and shapes. In such a world of unlimited storytellers, we will consume content in long and short forms, with words and with pictures and in what I call bytes, snacks, and meals.

When this happens, what's to stop CNN from creating an investigative report and selling it as an instant book with embedded video? Or Random House from selling a book with video interviews that are updated over time? Without the need for paper or disks, production and distribution costs will fall. Everything will become content that can be customized, combined, sliced, diced, pureed, and endlessly redistributed.

Some of this convergence is already apparent. CNN used to be a twenty-four-hour news outlet shown only on TV. The *New York Times* and the *Wall Street Journal* were simply newspapers. But on the Internet today, they are surprisingly similar. CNN's website has writers and editors, still photographs, extensive text, interactive graphics, and, of course, traditional videos. The *New York Times* and the *Wall Street Journal*, along with their traditional words, are offering embedded videos, interactive graphics, live interviews, and moving images. Online, the lines between television and newspapers have blurred—and soon the same will be said about books, movies, TV shows, and more. There is one more wrinkle: Amateur content and professional content are beginning to exist in unison, on the same devices with the same reach.

If all this makes your stomach feel uncomfortably queasy, you have plenty of company. Change as wrenching and new as this digital revolution in words and pictures is unsettling at best, rattling your security and bringing deep anxieties to the surface. It's true that business models and our traditional ways of thinking will have to change and that navigating that transition is difficult. But if it's any comfort, the advent of the printing press, trains, and television was similarly wrenching, yet we're much better off for having all of them.

If your main fear is that our ability to think deeply or focus on a subject is going to be washed away by the torrent of new information, relax. Even with this shift, long-form content isn't going to die. Kids may seem distracted, but they will play video games for an average of three hours a day—which sounds like long-form content to me. If they don't read a whole book in two days or stay with a television show, it isn't because they can't concentrate. It's because we haven't adapted the storytelling to fit their changing interests. They are consumnivores—collectively rummaging, consuming, distributing, and regurgitating content in byte-size, snack-size, and full-meal packages.

In this byte/snack/meal world, these consumnivores will drive the stories, deciding how much they want and what the format will be. If we want them to consume our stories, we'll have to harness a range of technologies to tell them well. If we don't, there are plenty of other options available for them to consume—or, more likely, they will create their next meal without us.

This Story

This book isn't about a list of absolute formulas for bringing in more revenue in a digital world. But for those of you wrestling with that challenge (or simply wanting to understand it better), this book will give you a new framework for looking at these difficult issues and making sense of the radical trends that have emerged in the last few years. I will take you deep into the consumnivore's new world, explaining how navigation, aggregation, and the narrative are changing.

To get a feel for the future as it exists now, we will go on a swing through the California porn industry, which through history has kept a step ahead of traditional outlets in trying new ideas and experimenting with the latest innovations in media. Then, to reassure you and put today's changes into perspective, we'll take a walk through history to see how radical new developments time and again have prompted fear and upheaval before proving their immense worth to society—and why we'll survive this sea change as well.

From there, I will lead us off the cliff into the shifting rivers, starting with our changing communities. Social networks, the openness of the Internet, and handy new devices are more than new ways to share photos, offer opinions, or waste time. As we struggle to make sense of the flood of information, gossip, and data gushing from the World Wide Web, these developing networks are providing crucial anchors that help us find our way. They help us determine what news and information we will trust and what we will ignore. As these new communities evolve and develop, they are profoundly changing how media outlets reach readers, how companies find customers, and even how we find and nurture our friends.

From there, I'll address the notion that our brains can't handle all this fast-paced stuff by diving into how these developing technologies are engaging our brains and how our brains are adapting to the volume of information flying at them from all directions. As part of that, I will take a closer look at one of the more successful of the current storytelling genres, video games, answering—once and for all, I hope—whether they're really bad for the next generation. As we all start to seek more compelling narratives and more engaging experiences, research in this field helps illustrate what the future of storytelling might look like. I will explore the needs of the next generation of consumers and creators who are at once creating and seeking new forms of narrative and immersive storytelling.

The next section can be summed up in one word: "me." The old role of media was to act as an intermediary between people and their understanding of industry, politics, and science. The job of media was to cull and curate for a broad audience. But consumnivores come to news from a different perspective: New technology has put each of them squarely on his or her own map, and now they want news that is highly personalized, relevant, and meaningful specifically to them. They are keenly aware that they and their friends no longer watch the same television shows at the same time and no longer will read the same newspapers or devour books in the same way. We are demanding that the stories of tomorrow be tailored to an audience of one—me—requiring a new approach. From there, I'll take you through the ever-growing debate about our compelling desire to multitask. We know we can't safely text and drive at the same time. But can the next generation of thinkers and consumers really chat, text, and still get their work done too? (The answer isn't as black-and-white as we've been led to believe.)

Finally, I will show you how the whole experience of consuming news, magazines, books, music, and other media is changing, and how the best morsels of information will stand apart from the voluminous clutter. This is the part where the old meets the new: Great storytelling, incisive reporting, and thoughtful editing will still prevail—but they will need to be presented to you and me in a different form to go beyond mere information. The people we buy content from must create a unique and meaningful experience for both communities and individuals and accept the fact that they will coexist with the amateur and the hyperpersonalized. I'll even look ahead ten years or more to see how today's cyborgs and 3D printers can show where we might be in a decade and help us navigate the ever-exciting world of tomorrow.

Speaking of tomorrow, you may wonder why I'm writing something as old-fashioned as a book to tell these stories about the future. Actually, this book is much more than the words you're reading here. Online and on your Web-enabled mobile phone, you will be able to mine a treasure trove of additional content. Some chapters will contain links to videos, visually walking you through research and new technologies. Other sections will link to extra information, including research papers, related news articles, graphics, and images. Additionally, as the Web allows today, you can go online to nickbilton.com and add to the discussion of each chapter through your social networks or with traditional comments.

As you will see, I eat my own dog food.

1

bunnies, markets, and the bottom line
porn leads the way

Oh, we're not going to wait [for the technology to exist to create content]. We're going to build it.

    —Ollie Joone, co-founder of The Digital Playground

I Did It for My Work. I Had To. Really!

Every second of every day, thirty thousand Americans type the word "sex" into an online search engine and hit enter. At least 50 million of our fellow citizens have done it. I've done it for a few minutes myself. Well, actually, for a number of hours.

There was a very good reason, though. I was doing research. Truthfully.

I did that research because the porn industry, unlike almost any other business, constantly has to try new approaches and new technologies to stay at least a couple of steps ahead of the morality sheriffs. It also must find fresh ways to satisfy the seemingly bottomless interests of its customers, who have been all too happy to move from well-lit arcades, to darkened movie theaters, to the privacy of televisions, to the very personal personal computer. As a result, the industry throughout history has been an innovator—and, over the last century, an early adopter of film, video, and the Internet.

So, I reasoned, the folks in the porn business should have some unusual and valuable insights into this shifting world of new technology, social networks, and free and paid content. To see if that was true, I had to check it out.

Of course, this would require vast amounts of research— hours upon hours of surfing the underbelly of the Web, looking at the best and worst of the porn websites. Honestly, I was trying to figure out who was making money online in this industry, though this intense exploration eliminated my ability to write or research from my local coffee shop, the *New York Times* offices, or any other public place. My wife, Danielle, was a little dubious too, to say the least. Eventually, she stopped asking what I was doing when stark nudity emanated from my computer screen and, at least for a time, tolerated my inquisitiveness.

It's a good thing she was patient. It took a little longer than expected to get to the heart of the industry. Although looking at pictures of naked people online is relatively easy, finding the real revenue in the industry that creates those pictures can be relatively difficult. Most adult companies are privately held, and though they revel in nudity, they keep their own financial matters well under wraps.

Thanks to the help of Lux Alptraum, a journalist and the

editor of the industry website Fleshbot.com (which, by the way, you should not investigate from your work cubicle), I was able to make contact with several players of various sizes in this under-the-covers industry. (Alptraum, who is spritely and in her late twenties, is always excited to talk about sex, porn, and the changing, blurring landscape of both topics. She understands the adult industry better than most journalists who cover it since she has been on both sides of the camera. Before she started writing about sex, she founded and ran a website called That Strange Girl, which was the first AltPorn site. AltPorn, she explains, is a form of online porn that shows "unconventional" models. Rather than the blond-haired perfect beauties you expect to see in glossy magazines, these sites feature people who look more like someone you would see on the street.)

As my quest continued, I made plans to head out to California, home of the film industry and most of the American pornography business. The industry has thrived in the Golden State for two reasons: First, there seems to be a lot of "talent," partially because of the traditional movie industry. Second, California has a lax legal climate compared with other states, where people who videotape sex can be charged with any number of illegal acts, including pimping.

California wasn't always lenient. In 1988, the state accused Harold Freeman, a porn creator, of being a pimp as part of an effort to clean up and shut down the porn industry. In *California v. Freeman*, the state likened the act of taping and selling porn to that of prostitutes selling sex on the streets. The case lasted several years and made trips to both the state

court and the U.S. Supreme Court before a ruling finally set-
tled that creating and selling porn was different from selling
actual sexual acts.

As I set up interviews, one company spokeswoman asked
who I would like to interview beyond the top executive. Would
I like to meet and talk with the "talent"—that is, the film stars?

"Maybe you would like to interview Jesse Jane, Stoya, or
we could even try and get you an interview with Tera Patrick,"
offered the spokeswoman.

Oh, intriguing, I thought. I told her I'd get back to her.

Over dinner that night with friends, I told Danielle about
the enticing offer.

"You don't need to meet with porn stars," she said.

"Well, maybe I do . . . for the book."

"No, you don't," she said firmly.

So much for tolerance and understanding.

Porn, like its subject matter, is always eager to
experiment.

—Peter Johnson, "Pornography Drives Technology: Why Not to
Censor the Internet"

Before we gaze into the future, it might help to peer into
history. Adult content—pornography, that is—has roots that
can be traced back thousands of years. The ancient Greeks displayed risqué statues and paintings in market squares; the his-
torical artwork and sculptures discovered in Pompeii include
an array of lewd and phallic paintings and statues.

In a mid-1990s essay about the Internet and censorship,

lawyer Peter Johnson's assessment was: "Throughout the history of new media, from vernacular speech to movable type, to photography, to paperback books to videotape to cable and pay-TV to '900' phone lines to the French Minitel, to the Internet to CD-ROMs and laser discs, pornography has shown technology the way." He quotes Camille Paglia, a self-described dissident feminist, who said that "great art is always flanked by its dark sisters, blasphemy and pornography." Johnson went on to point out that "the same is true of the mundane arts we call media."

Johnson noted that Chaucer's *Canterbury Tales*, which first appeared in the mid-1300s, was "larded with sexy and scatological" content—and was both sought out by the cultural reading elite of the day and "read aloud to a largely illiterate populace," helping create the vernacular language of England.

Once the printing press debuted, the Bible became popular reading, but it had some competition from some more colorful offerings. *Sixteen Postures* by Pietro Aretino was a series of engravings of sexual positions, and *Gargantua and Pantagruel* by François Rabelais in the sixteenth century included stories and etchings of sexual encounters that were widely distributed throughout Europe. Rabelais, a famous French writer, boasted that more of his sexually explicit books were sold in two months than copies of the Bible were sold in years—although since BookScan, the database that tracks book sales, wasn't developed until the twenty-first century, official figures aren't available to prove it. He did, however, offer prescient advice to those in the media business: Sex sells.

Centuries later, the roots of and eventual birth of early movie theaters grew out of early movie arcades, where a person could insert a coin and see a short, fuzzy clip of a woman undressing. The clips were quite modest by today's standards and just a few minutes long, but they were enticing enough that customers would continue to add coins to see where the picture would go. Before viewers knew what they had done, ten to twelve scenes had passed and they unwittingly had paid for a short movie. In San Francisco in the early days of such movie arcades, it was estimated that they brought in millions of dollars in revenue.

In the 1970s, the desire for pornography played a role in helping settle a long-running battle for the technology that would drive the VCR that used to sit in your living room under your television. (For some of you, it might still sit there.)

Betamax was developed by Sony. Its competitor—VHS— was developed by JVC. Betamax was superior in quality, but the design of the tapes limited the length of videos to one hour, making them perfect for recording television shows. VHS tapes, by contrast, could run up to two hours, making them much better suited for movies.

For about a decade, consumers were stuck in the middle of a rivalry over these competing technologies. In the end, although the Betamax was arguably a better product, VHS won the tape wars and Beta disappeared. One of the factors can be traced back to Sony's stance on adult content and the strict antiporn policy it implemented for the usage terms of its tapes. This policy prevented any adult content companies from using or distributing pornographic content in the Betamax format. Adult film producers had no choice but to use VHS, and in turn a swath of early technology adopters and porn consumers purchased more VHS VCRs and fewer Betamax systems.

The adult industry also found income from the telephone. After a U.S. Justice Department antitrust suit led to the breakup of AT&T in 1982, "Ma Bell" was split into several operating companies, bringing competition to the telephone industry. The adult industry found a new way to make talk expensive. For decades, the telephone had been used for local and long-distance conversation between friends, family members, and business associates. But with pay-by-the-minute 900 numbers, porn purveyors found that people would pay a fair amount of money to talk to someone with a name like Sparkle, Mercedes, or Bruce about anything that caught their fancy. This paved the way for others to charge for horoscopes, business and legal advice, and even the weather, delivered by phone. Surprisingly, these pay-for-porn phone numbers still exist around the world and are still used widely in Europe and Asia. Although the pay-for-legal-advice hotlines didn't take off, the computer and help-desk industry took advantage of this unique business model. Today you can call expensive customer support numbers to get help with your computer. But once again, the pornography industry was an early leader.

Then there were the early days of the Internet—a land of science papers, message boards, and porn. Many of the first porn images online were scanned by early Web users from magazines and posted to Usenet message boards.

As the audience of the Web grew, so did the amount of sex-related content available online. By the mid-1990s, many porn sites were bringing in millions of dollars while many mainstream websites were struggling to figure out how to make any money online. Though photos and videos might take minutes to appear over telephone dial-up lines, porn peddlers were doing a brisk business in images and video. Again leading the way in new media models, they were among the first services to charge successfully for online subscriptions and use encryption for credit-card payments.

All of that led me to think that today's porn purveyors—the test drivers for new media—might have some insights for the rest of us. Was there, in fact, a business model for content and storytelling? I assumed that if the porn industry could solve the problems and transitions from print magazines and analog DVDs, surely the rest of the media industry could follow suit and save itself from imminent death. After all, I thought, the porn industry has done this so many times before. Maybe it had solved today's technological transition too.

I was excited to venture out into the world of porn. (No, not in that way!) I enthusiastically and optimistically assumed that the *Playboy*s and *Penthouse*s of the world had figured out new business models: how to charge for content and how to tell stories in this new digital age. This trip would be a no-brainer, I thought. I'd return to New York with the secrets of the future of media like a time traveler carrying a winning lottery ticket.

Despair

It took only a few meetings before I realized that the lottery ticket didn't exist—or at least that's what I was told. Despite my excitement about discovering new ideas in the California porn industry, I heard mostly fear and desperation from directors and the people who run production houses. Prices were falling. The barriers to entry had disappeared. Some money was flowing in, but advertising and sales of traditional media were in decline and it wasn't clear how long even the current business could sustain itself. The industry, I was told, was being attacked by parasitic piracy and file sharing.

"It's tough now," said one porn purveyor. "We're dealing with piracy, and we're dealing with free content on the Internet, which is eroding the normal business models that have been around for many years, like DVDs, magazines."

"We tried to fight piracy as much as possible because it is hurting us," said another marketing employee.

"Well, we're making money now from online sales, but I'm not sure how long we can keep this up," a company owner said.

A successful producer railed for twenty minutes about the death of the adult industry. Once he had to worry only that a competitor would have a hotter star, a sexier scene, or better distribution. But now his competition is everywhere, so vast that he can't put his arms around it, and there's no way to stop it or slow it down. Any eighteen-year-old with a $300 netbook, a Web camera, a $25-a-month Internet connection, and a Pay-Pal account can make live nudity available to anyone who is willing to pay for it. And that doesn't even count the ocean of content available for free. Given that, how was he supposed to pay employees, taxes, office rent, and other bills?

The paradox is that porn is still popular—maybe as popular as it has ever been. An estimated 36 percent of all Internet users check out at least one adult website each month, according to comScore, which tracks websites and Web users. In 2008, the entire pornography industry brought in estimated annual revenue of around $20 billion, a number that grows a bit each year. Figures collected by AVN Media Network, an adult industry reporting news group, indicate that consumption online grows about 13 percent each year and in 2006 was responsible for a total of $2.8 billion, or about 14 percent of all adult-content revenue. The industry has also seen healthy and consistent increases in pay-per-view cable, merchandising (sex toys and other erotic paraphernalia), and of course mobile websites and applications for mobile phones.

But as with traditional media—books, newspapers, magazines, movies, and the like—the traditional purveyors of big breasts and hot poses are shrinking as their best customers turn elsewhere. Sales of adult magazines are falling 5 percent on average each year, and video sales and rentals are falling a dramatic 15.4 percent annually. Walk into any adult video store in America and you'll see DVDs that were supposed to sell for $50 discounted to $5 or $10.

Then there's the unraveling of the iconic Playboy Enterprises Inc., a rare publicly traded company in the industry. Between 2004 and 2007, *Playboy*'s revenue was between roughly $330 million and $340 million and the company turned a small profit or basically broke even. But in 2009, revenue slid to just $240 million, a drop of $100 million—almost 30 percent—in just two years as television, video, and print results plunged amid a technology shift and an economic recession. Losses totaled more than $50 million. The company's stock, which had begun the new century trading above $25 a share, began 2010 trading at less than $5. Its outlook wasn't any more encouraging: In late 2009, the company said it would print one fewer issue of the magazine in 2010.

A senior-level *Playboy* manager confided that the company had become mired in bureaucracy and organization charts and had tried to innovate by committee. In meetings, managers didn't talk about "how can we be ready for what's next" but instead were fixated on "how can we continue to get people to buy our DVDs and magazines."

How desperate was the company? To counter dwindling revenue, it was reproducing its logo, well, like bunnies. The *Wall Street Journal* reported that *Playboy* had "turned its bunny loose" with a range of perplexing and seemingly desperate licensing efforts. Among other things, it was "slapping its famous logo on a tanning spray, a disposable lighter, a mattress, a couch, and a line of drinks designed to boost the libido." So much oddball stuff featured the bunny that even diehard collectors weren't interested.

From interviews and research, it appears that the *Playboy*s and *Penthouse*s of the world believed that their downturn was just an economic hurricane and that they would be able to rebuild and return to normal once the winds died down and the storms passed. Not only is this proving to be wildly optimistic, it's also proving to be their demise.

If this sounds familiar, it's because so many other industries—newspapers, books, music, and movies—feel like they've been wrestled to the mat by very similar issues. Traditional products that rely on expensive advertising or products sold over a counter still pay the bills and keep the lights on, and they can't figure out what will replace that and how. So instead of simply capitulating to some consumers' demand for a new approach, they are, understandably, trying to hang on to as much of their revenue as they can and trying to persuade their customers to stay with their tangible goods while they experiment with new technologies and scan the landscape for an answer.

Justifiably, there is plenty of despair, even in the porn industry. I'm usually perversely optimistic when it comes to technology. But after a week of hearing that the sky was falling, I have to admit that I wasn't too confident in the future.

In the middle of my travels through the porn studios of California, my optimistic vision of the future of media, based on the tribulations I had heard through the week, didn't seem so bright. Over and over I had heard stories of despair and defeat. I heard about rampant piracy and the free content people were creating in their bedrooms with cheap webcams. I heard that people no longer wanted DVDs or magazines, which was no surprise, but in addition, they weren't willing to pay the same price for online content.

I too believed the despair—maybe the sky had already fallen. If the pornography industry, a trade that had been tested for hundreds of years, couldn't figure it out, maybe newspapers, magazines, movie houses, and everyone else that sold content for a living should just throw in the towel.

The rest of the week didn't go much better. But on the plane ride home to New York, as I sat rustling through my notes from interviews, I saw things a little differently. From a pile of interviews with companies big and small, niche and mainstream, I came to see something else. Yes, it was true that the porn industry didn't have a single "aha" answer to the new age of content creation and consumption, but collectively, it had numerous answers. In many ways, a new industry is being built from the ruins of the crumbling legacy companies. Collectively,

those experiences might help explain what future content outlets will look like and provide lessons in how to adapt.

From Perfect Bunnies to . . . ?

The average *Playboy* bunny over the last fifty years is blond,
blue-eyed, 21.7 years old, and five-feet, six inches tall and
weighs about 115 pounds. Maybe she was the dream of every
man in a different generation. But no more.

Jo Mason was once a high school principal. She's tall and
confident and speaks with a calm understanding that would remind you of a caring and tolerant aunt—someone you can talk
to about anything. She looks like someone who promises not
to judge, and doesn't.

She ended up in the porn industry almost by accident. Several years ago, a friend who ran some porn websites needed
some help on a project, and Jo agreed to lend a hand on a temporary basis. Before long, she was running X-rated websites
in her free time while keeping teenagers in line during the day.
Eventually, word started to get around about her moonlighting,
and she had to make a choice.

She chose the porn business.

"I've used all the same skills in both businesses," Mason
said. "There was a natural continuation and connection, and
there really wasn't much of a transition between managing high
school students and young and coming porn stars."

Mason runs a few small porn sites and compares the giant
adult-content companies to the legacy car companies such as
General Motors and Chrysler that crashed into bankruptcy

proceedings in recent years. Those companies refused to innovate, refused to give up on big sport-utility vehicles even as gasoline prices climbed and customers clamored for fuel efficiency and hybrids. Whereas the legacy adult-content companies pushed magazines and videos, Mason and other operators of small sites offer customers niche content. One of her sites, for instance, is called Little Mutt, featuring new unknown models—usually in their early twenties—engaging in straight and lesbian sex. Although similar content can be found free online, Mason tries to post higher-quality content with better lighting and production values—in other words, professionally produced porn with an amateur-like appeal.

Although the top-heavy, blond-haired, blue-eyed beauty may still be an ideal to some men, customers of adult content are showing they want a more personalized experience that fits their specific tastes. The Internet, which doesn't need fancy equipment or distribution agreements, will provide it. Maybe customers are interested in black women, Latinas, Asians, striped stockings, older women, bigger bottoms, smaller tops, or some quirky combination. The Internet, which has no commitment issues, will provide it. A universal brand is not enough for today's consumers. If consumers want gritty, that's what the Internet will give them.

Turns out, customers will also pay for it. Benjamin Edelman, a professor at Harvard Business School, has explored how consumers decide which adult sites they are willing to pay for. The graph on page 33 shows a price index of monthly online subscription fees for online adult content.

You can see that there's a strict limit to how much people will pay for porn. Customers are willing to spend $5 to $25

FIGURE 1

Subscription Prices
(in dollars for subscriptions of one-month duration)

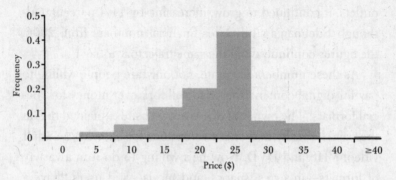

Source: Analysis of reviews from Rabbit's Reviews.
Based on a graph published in the *Journal of Economic Perspectives* (Vol. 23, No. 1, Winter 2009), published by the American Economic Association.

a month for a subscription to an adult website. But after the monthly subscription price passes $30, willingness to pay drops off a cliff. In other words, consumers will pay for specific porn that they want online, even in the face of free content, but there is a threshold to how much they will spend even for obscure, niche content.

Even with such low subscription prices, websites can make a profit because they started out with low overhead and few employees.

Edelman's study also foreshadowed the reduced income organizations are facing as consumers move away from products such as videotapes, DVDs, and magazines in favor of digital experiences on mobile phones and computers. Citing industry statistics from AVN, Edelman shows that video sales and rentals fell 15 percent between 2005 and 2006. Digital, in contrast and not surprisingly, grew across all forms of delivery. Internet sales grew 13.6 percent in the same period, and although mobile porn was still small in comparison to other outlets, it continued to grow, increasing by 11.4 percent. (Although Edelman's study cites financial numbers from 2006, the figures continue along the same trajectory today.)

As these numbers illustrate, not only are people willing to pay for digital content, they also will fork over money for special formats. The owner of one video website explained that in the past, he really had only two options for showing an adult video: VHS and DVD. Now he's willing to do it in a variety of formats—in bytes, snacks, and meals—and users "have a choice with how they want to view it." So if someone wants to watch a thirty-second clip on a mobile phone, he'll sell it to that person—just as arcade owners sold short bits decades ago. Does the customer want a feature-length high-definition DVD? Step right up; he'll be glad to provide whatever people are willing to pay for.

These pornography sites charge any number of different prices for their content. But these companies realized that they have to make the content that consumers want, and they have to make it available anywhere, at reasonable prices, any time the consumer wants it. Most important, since production costs and distribution channels no longer create a barrier to entry, if these companies don't do that, someone else can and will.

To bring in revenue, the smaller sites have recognized that advertising can be enough to pay the bills and keep the lights on. But the ads must be relevant to their audience. So if viewers see an advertisement similar and relevant to a clip they are about to watch, there's a good chance they'll click through to the ad's link. But if the consumer is watching porn and the advertisement is for a car, the ad probably won't generate many clicks.

Catering to consumers' specific preferences also helps fight the wave of free or stolen material online, the plague that is causing so much angst and frustration in the news and mainstream entertainment businesses. This digital shoplifting dates back to the dawn of adult content on the Web but has expanded over the last few years. For instance, a bevy of websites called tube sites—the porn versions of YouTube where anyone can upload and download content—have sprung up to post ripped-off or user-submitted content under names like Youporn, Red Tube, and Xtube. Just as ABC, CBS, and Viacom must confront having their content posted by viewers on YouTube and other online video outlets, the pornography industry has been forced to deal with this illegal content sharing too.

Although some video sites have tried to shut the tube sites down, other porn purveyors have taken a different approach, says Alptraum, the editor of Fleshbot. Rather than spend tens of thousands of dollars in legal fees trying to shut down tube sites—money that many smaller sites don't have—the content makers decided to adopt the old mantra "If you can't beat 'em, join 'em."

Producers started uploading teaser versions of their own content on the free video sites. They wanted to create experiences that would lure a user from a tube site to their own sites, where more content—and related advertising or sales offers—awaited. To do this they took two approaches: The first was sharing new, original content that didn't already exist on a DVD—content that couldn't have been illegally copied and posted yet. It was sort of like offering a free toy with a Happy Meal at McDonald's: When all is said and done, the consumer isn't sure if the toy or the food is free—but it doesn't really matter as long as McDonald's made a sale.

The second approach involved upping the stakes. If someone uploaded an illegal version of a video to a tube site, some content owners would upload their own version in the form of a higher-quality clip, only slightly longer, with links and ads embedded to help bring the viewers back to their home pages. In many instances this has worked. Go to a tube site today and you'll see high-quality video uploaded by porn creators sitting next to stolen content that's a little old and grainy. Which one are you going to click on?

After my California trip, I went back to Alptraum to share my findings. She invited me to her office so that she could show me the results of a survey she had recently given to her readers.

Fleshbot, where Alptraum is the editor, is a part of a much larger company called Gawker Media, which is the parent company to several well-known blogs. Gawker was started by Nick Denton, a journalist turned entrepreneur who began the company in 2002 with a technology blog called Gizmodo. At that time, "blog" was still a very insider-technology term. It's true today that almost everyone has a blog, even the White House. The *New York Times* has several—and I work for one of them! But in 2002, blogs were sparse and were considered more of an online diary than a viable business. When I asked Denton his reasoning in starting the website, he answered with a very logical response.

"One day I was reading *Wired* magazine," he explained, "and I thought to myself, Why does this publication only come out once a month? Why can't it happen all the time—maybe even every hour or every few minutes?" Gizmodo is now one of the biggest technology gadget blogs on the Internet and draws in more than 150 million page views a month. After the success of Gizmodo, Denton decided to expand. More blogs were launched off the Gizmodo concept, including the well-known gossip blog Gawker and a variety of other sites. In a similar fashion to the porn industry, Denton realized that customers want niche products. Collectively, the Gawker Media sites generate nearly 400 million page views a month, and all of them are free, able to deliver niche advertising to a niche audience. Denton was able to grow the sites so quickly and become almost instantly profitable partly because he didn't have to battle the forces of a bricks-and-mortar business. There are no printing presses or distribution problems to deal with. Instead, the blogs' writers are paid by the clicks on their stories and can work from anywhere. (Most work from home.) A handful of editors, including Alptraum, work in an office in New York City.

The Gawker Media offices are housed in an old garment building in an area of the city called NoHo, occupying a floor with deep-red brick walls and rickety wooden floors. The offices are set up in a way that reminds me of a supermarket with long aisles. But instead of dairy products and cereal lining the shelves, prolific young bloggers sit at rows of desks in front of computer screens, typing away and serving up content by the pound.

The receptionist pointed me to Alptraum's desk at the far end of the room. As I wandered in that direction past each blogger's desk, I glanced at computer screens displaying different niche topics. One person was looking at pictures of a souped-up truck for the car blog Jalopnik. Another played with some gadgets, probably for the tech blog Gizmodo. At the next desk, someone was editing images of a video game, maybe a writer for the gaming blog Kotaku. Finally, I got to Alptraum's desk, where, as you might imagine, her screen was covered with pictures of naked people—specifically, a video of two people having sex.

Alptraum looked up without making any attempt to hide the screen and said, "Hey, Nick! Great to see you! Just give me a second. We just got this new celebrity sex video, and I want to get it up on the site." I watched as she bounced back and forth between her Web browser windows and quickly published the post.

When she was done, I asked if she ever felt uncomfortable looking at porn all day at work. "No," she replied. "It's my job, and I don't really even think about it as porn anymore. I think of it as providing content for an audience.

"Sure, my screen is filled with penises and tits," she continued, "but that doesn't mean my job is any different than the guy over there writing about video games or gadgets. It's just niche content that people are interested in."

She told me about a survey she had just asked her readers to fill out on the Fleshbot blog: "Porn Worth Paying For: What Makes You Open Your Wallet?" http://fleshbot.com/5318653/porn-worth-paying-for-what-makes-you-open-your-wallet. The readers' responses were split into two camps. Some said they would love to pay for porn online but the prices were still too high. "I refuse to pay more than $15 for any porn DVD," wrote one reader. Another said, "I was going to buy my first porn DVD just last week, looked at the prices and laughed and went to download instead."

But most of the readers said they would pay for quality or storytelling. "I tend to pay for decent 'plot driven' feature types more than anything," wrote one reader. "I'm all about the niche content. Those are the people I want to give my money to," wrote another. "Well-crafted feature porn is far more enjoyable to me, and even worth paying for," yet another said. "I'm happy to pay for a quality website that's full of great content. I'm seriously considering getting with plumperpass.com," another confident reader wrote.

Even in the down and dirty world of smut, quality matters. "We can see people have been online for years paying for niche, quality content, and interaction. Those numbers exist," Alptraum said. "As long as the price is right and the content is professionally shot and offered in any number of formats, people will pay."

In other words, people will pay for well-packaged offerings—even in the face of free alternatives.

Still, Alptraum cautioned, this is not always the case. "There are some instances where people are happy just getting off to a really poorly shot free sex video—even sometimes shot from a shaking grainy cell-phone camera," she said. "But for most people, even when it comes to porn, quality will always be worth paying for"—though she added sternly, "as long as the price is right."

But porn companies that have tried to charge too much have seen their content stolen and shared all over the Web. "The right price, quality, niche, and immediacy," Alptraum reiterated; "that's what people will pay for."

Experience Matters

In my travels through the porn industry, it was evident that the smaller start-up companies are innovating and pushing the boundaries of the medium. They are listening to their customers and creating content that customers are willing to pay for and delivering it to the devices where they want to enjoy it.

Some porn companies recognize that today's customers are also consumnivores—in one form or another, we're all consumnivores, especially the next generation. We're constantly cutting up content, picking out the best pieces, and passing it along. In the past, my mother used to do something similar, but on a much smaller scale. She would grab a pair of scissors and snip interesting articles from the local newspaper or recipes she wanted to try from a magazine. Now a generation exists with the mentality that those scissors have been replaced with a mouse and an Internet connection. And whereas my mother used to cut out entire articles from the paper, the equivalent today is to slice and dice words, images, paragraphs, and video clips. The audience doesn't necessarily need to pay for someone to do that for them.

But there's one other thing I discovered the next-generation consumer will pay for online: better experiences, which often grow out of better storytelling.

Sometimes that takes the form of relationships—not in the sexual sense but in the way you connect with your consumers and create new communities.

For more than a decade, well before sites such as Twitter, Facebook, and Friendster existed, some players in the adult industry engaged in their own version of social media. They didn't really know what they were doing, and no label was applied to their practice. They simply recognized the importance of developing a connection and community with their audiences.

In the late 1990s, when niche pornography sites began popping up all over the Web, some adult actors and actresses began going on message boards on their websites and chit-chatting online with the customers who paid for their content. Sometimes they would describe a scene they were going to shoot or even share their plans for that evening. They tried to engage in one-on-one discussions with the customers and in doing so created the bond many are trying to create today with social media websites such as Twitter and Facebook. In those early days, they recognized the importance of conversation.

That proved to be somewhat of a turning point. There are many reasons people steal content, as I'll discuss later. But one of the major problems of the Web is the lack of humanization. People are oblivious to the fact that a human being is on the other side of the digital information they are consuming. The people who copy adult DVDs and upload them to tube sites don't consider that a human being might be making a living from that content. But 99 percent of those uploading the footage would never walk into a XXX store and steal the actual DVD.

By going into these forums and sharing their personal stories with people who had access to their content, the porn stars added a dose of humanity and community to their digital images—a very difficult task online but one that is being introduced slowly by mainstream publishers with the adoption of social networks. Once visitors engaged in conversations on the porn sites, many of them no longer felt comfortable stealing and sharing the work of people who were trying to make a living. They simply saw them in a different light.

Personal stories add one dimension, but great storytelling on the screen or the page consistently stands out. Yes, it's true that the adult industry will face competition from some people performing in front of a webcam in a bedroom or using a mobile phone connected to the Web. The mainstream media are going to suffer the same fate too. What's to stop someone from writing a blog post about a breaking news event because she finds it interesting or reviewing a restaurant he enjoyed? Nothing. And as has happened with the porn industry, the next generation of content and media will exist in the same way: professional sitting alongside amateur. Although better content and better stories nearly always trump amateur hour, they clearly will coexist next to each other in the future—just as porn content does today on the Web.

But the porn industry shows us that people will pay for good storytelling. Ollie Joone understands this better than most in the adult industry. Joone entered the porn world in 1993, well before the Internet was a household necessity, and began making adult CDs. The company he cofounded is called Digital Playground and claims to have 40 percent of the adult video market, providing skin flicks for hotels, cable, and pay-per-view TV. Joone says adult films aren't just about selling sex; they're also about storytelling and the overall experience. The company still uses big-name porn stars and has built part of its business with clever, sexy spoofs of popular films, such as *Pirates,* a play on *Pirates of the Caribbean.* The *Pirates* version, a multimillion-dollar production that was actually shot on a ship, won an R rating and made millions of dollars in video sales. *Pirates II* is now in the works.

Digital Playground's tagline is "porn worth paying for." I asked Joone how he differentiates his work from a quick clip of someone naked. He explained it like this:

Imagine you're watching a movie with a dramatic car chase. If it was a really good chase, with police cars and sirens, the quality of the video almost wouldn't matter. Just the content by itself would be dramatic. Now imagine that you know the backstory of this chase, that it's a life-and-death matter, that someone's been shot—maybe they just robbed a bank—or that one of the cop cars is stolen too. It would make the video experience much more compelling. Add a heightened level of quality and interaction to this and you've got an experience people will pay for. It's the same exact mentality with pornography, he said.

The day I interviewed Joone, he was off to shoot a scene for a fictitious sex party, using a new technology that would make the action look three-dimensional. Using equipment that allowed up to a dozen cameras to record the action at the same time from different angles, he could produce a picture that would allow movie watchers to look toward any direction in the room, see the scene from multiple angles, and feel almost a part of the action, much like the experience of a very good video game.

As I wrapped it up, I asked Joone what was next for his industry. Technology isn't available yet to do what he'd like to do, he said. But he believes the next generation of porn and storytelling will be hyperpersonalized, placing you almost directly in the scene. That will give you control over what you see—almost like you're standing on a holodeck, a room that uses holographs to simulate reality.

How long will the industry have to wait before the technology exists and he can start to create content like this?

"Oh, we're not going to wait," he quickly replied. "We're going to build it."

Looks like the porn industry is leading the way after all.

2

scribbling monks and comic books

it's ok—you've survived this before

Thus the telephone, by bringing music and ministers into every house, will empty the concert halls and the churches....

—"The Telephone," *New York Times*, March 22, 1876

The world has been going to hell for a long, long time. So if you're feeling rattled by the stunning growth of today's new social media and are afraid that the way humans communicate is going to change rapidly—and not in a good way—your fears are understandable. Time and again, new technologies have been seen as frightening, intimidating, and a sure road to ruin.

We shudder at the unknown. We know deep in our hearts—sometimes quite correctly—that the world is about to be screwed up in the name of progress. New developments often seemed poised to ruin a perfectly good way of life. At various times, they have seemed dangerous (or even life-threatening), destined to destroy our personal relationships, or poisonous to our culture, our language, or our basic manners.

Yet we're still here. Despite the hand-wringing of the *New York Times,* we still go to concerts and lectures, even though the much less expensive option of enjoying music and speeches is readily available on our superthin iPods.

That possibility seemed unimaginable to the newspaper in 1876, when it wrote about the potential impact of the research of a Professor Reuss. "A distinguished German performer on telegraphic instruments who has recently made an invention which cannot fail to prove of great interest to musicians and, indeed, to the general public," the paper said. "The telephone—for that is the name of the new invention—is intended to convey sounds from one place to another over ordinary telegraph wires, and can be used to transmit the uproar of a Wagnerian orchestra or the gentle cooing of a female lecturer." That seemed to be a good thing and certainly convenient. But there was a dark side:

"No one who can sit in his study with his telephone by his side and thus listen to the performance of an opera at the Academy will care to go to Fourteenth Street and to spend the evening in a hot and crowded building. . . . The rural visitor who spends a Sunday in town and reads a printed notice in the office of his hotel to the effect that 'Talmage's sermons . . . can be had at eleven o'clock in the telephonic room,' will, of course, give up his original intention of risking a journey to Brooklyn. . . . Thus the telephone, by bringing music and ministers into every home, will empty the concert halls and the churches. . . . It is an unpleasant task to point out a possibly sinister purpose on the part of an inventor of conceded genius and ostensibly benevolent intentions. Nevertheless, a patriotic regard for the success of our approaching Centennial celebration renders it necessary to warn the managers of the Philadelphia exhibition that the telephone may really be a device of the enemies of the Republic."

But before Reuss (whose name actually was spelled Reis) had a chance to destroy society as it was then known, along came Alexander Graham Bell's version of the telephone, which for many decades has allowed us not only to be in constant contact with friends and loved ones but to perform business transactions from thousands of miles away. Although the *Times* noted that the telephone could bring others' voices into the home, the author was fearful of the uncertain future—and sure it would eliminate the need for people ever to leave the house. People were clearly scared of the possibilities, but it didn't take long before another technology reared its ugly head.

Just a year and a half later, the *Times* was even more despondent about the phonograph, which could preserve those precious sounds and words for years or decades to come. "The lecturer will no longer require his audience to meet him in a public hall, but will sell his lectures in quart bottles, at fifty cents each; and the politician, instead of howling himself hoarse on the platform, will have a pint of his best speech put into the hands of each of his constituents," the paper wrote in November 1877.

But the real danger—the most serious threat to society—lurked ahead, the paper warned: "There is good reason to believe that if the phonograph proves to be what its inventor claims that it is, both book-making and reading will fall into disuse. Why should we print a speech when it can be bottled, and why should we learn to read when, if some skilled elocutionist merely repeats one of 'George Eliot's' novels aloud in the presence of a phonograph, we can subsequently listen to it without taking the slightest trouble? . . .

"Blessed will be the lot of the small boy of the future. He will never have to learn his letters or to wrestle with the spelling book. . . ."

Fear of the new and fear of the unknown are common afflictions. At their worst, they can stunt or stop innovation. More commonly, though, this technology hypochondria—or technochondria, if you will—rattles a large part of the population, leading to a divide between those who rush forward with new experiences, fearful that they might miss something, and those whose fright leaves them feeling disoriented and left behind.

With so much anxiety, it can be hard, if not impossible, to board the moving train—quite literally. Even the arrival of train transportation came with a railcar full of fears that left some holding on tightly to their horses. A number of historians note that the railway brought an incredible amount of anxiety across all levels of society. For example, according to one history book, the nineteenth-century beginnings of rail transportation in Great Britain stirred up "extraordinary paranoia": "It was claimed that trains would blight crops with their smoke and terrify livestock with their noise, that people would asphyxiate if carried at speeds of more than twenty miles per hour, and that hundreds would yearly die beneath locomotive wheels or in fires and boiler explosions. Many saw the railway as a threat to the social order, allowing the lower classes to travel too freely, weakening moral standards and dissolving the traditional bonds of community."

That's right: Some people theorized that if humans traveled

at more than twenty miles per hour, they would suffocate. Or worse. Anne Harrington, chair of Harvard's history of science department, found that scientists also believed that traveling at a certain speed "could actually make our bones fall apart."

After reading numerous articles, papers, and discussions of the mid-1800s, Harrington discovered that nerve specialists and psychiatrists, including extremely well-respected scientists and physicians, were behind those theories too. Eventually these medical conditions earned their own diagnoses. Nineteenth-century citizens suffered from such ailments as "railway phobia" and "railway spine," a debilitating result of sudden stops. This wasn't an affliction to be taken lightly. In 1867, John Eric Erichsen, a well-respected fellow and professor of surgery in Philadelphia, wrote one of the many books on the topic, titled *On Railway and Other Injuries of the Nervous System.*

In a normal progression, fear of the new morphed over time into a fear of the unknown outcome. "The more extreme fears did recede as the railways spread, becoming established as an economic and social necessity and proving their ability to function safely and reliably; yet below the superficial acceptance deep disquiet remained," the railway history continued. "Rather than disappearing altogether, the fear and anxiety provoked by the railway changed in nature as the nineteenth century progressed, becoming a fear of internal rather than external disruption.

"The reasons for this change lay in the unique potency of the railway as a symbol of modernity. In the scale and sophistication of its engineering, the order and complexity of its operation, the speed and power of its technology, the railway embodied all the forces of mechanization, organization and industrialized progress which lay behind modern civilization." Like many technologies today, its true long-term impact was hard to gauge.

From 122 Books to 7 Million

Fear certainly makes for good headlines. But fearful and anxious reactions to innovation also keep us from seeing the bigger potential of new ideas. There's an all too human tendency to believe that what we know and experience now is the way it will and always should be.

So in worrying that the telephone and the phonograph would replace concerts and reading, critics of the day were simply not able to perceive the possibility that those devices would bring music and ideas to a much broader audience. Most people could not envision that phonographs—followed by cassettes, followed by digital downloads of music—would build such a base of fans that some day a hundred thousand people would gather to hear live concerts at a football stadium.

The printing press was subject to the same kind of narrow thinking. When Johannes Gutenberg used his revolutionary invention to publish the Gutenberg Bible in 1452, his work didn't make much of a splash. Until then, books had been painstakingly copied by hand by monks. Each letter was intricately drawn, each word planned out, mulled over, and painstakingly transcribed. Making books was considered an art form—actually referred to as the "black art." (This line of work received its ominous name from the black ink that stained the workers' hands after a long day creating type.) Readers, for the most part, were scholars and the religious elite.

If you traveled back in time to 1424 and entered the University of Cambridge in England, you would find one of the largest libraries in Europe. Here you could see an impressive list of 122 books. The books were delicate, large, and beautiful. As the books were made by hand, it took another fifty years before the collection would reach an even more admirable 330 books. (Today, the University of Cambridge has more than 7 million books.)

Then, out of the blue, what had taken monks months or even years to produce could be done in a matter of hours. As word of the printing press gradually spread through Europe, the monks were curious about the new technology but didn't see any reason to worry about it. To them, such a modest reproduction couldn't hold a candle to their exquisite handcrafted works. In addition, most laypeople were not literate, so the new technology essentially was being tested in a vacuum. Most people in fifteenth-century Europe weren't interested in books and wouldn't have cared about what a printing press could do. So even as the presses started to gain traction in the bookmaking industry, many dismissed the new technology. Those who wrote books by hand simply considered the product inferior— until it had largely replaced their trade.

Some politicians and clergy, however, despised this innovation. As Elizabeth Eisenstein chronicles in *The Printing Press as an Agent of Change,* the presses were the basis for the artistic Renaissance, the religious reformations, and the scientific revolution that spread new perspectives about physics, anatomy, and a variety of other sciences. Those powerful ideas helped move society from the Middle Ages to the modern sciences, displacing the ideas of the Church.

The printing press allowed the spread of information that couldn't be controlled by the clergy, kings, politicians, or the religious elite.

Still, it took a while for books to evolve into something that could be shared easily. Early books handmade by monks were massive and monstrously heavy, sometimes weighing more than fifty pounds, and about the width and height of a newspaper today. They weren't the least bit portable. If you wanted to read a book, you went somewhere to do it. You certainly didn't take it with you.

When Gutenberg and his collaborators developed the printing press, their goal wasn't to innovate a new size or shape; it was to innovate the speed of production. The Gutenberg Bible, for example, consisted of two volumes and 1,286 pages. It was so heavy that it could be read only when one was standing at a lectern.

According to the book historian Alistair McCleery, it wasn't until 1502 that Aldus Manutius in Venice came up with smaller, more portable books "that did not require a lectern or reading stand, or cause the reader's arms to ache from holding them." Manutius essentially invented the mobile phone of his day. He came up with the idea of smaller, mobile books that people could carry around and read anywhere—the first portable books that could fit in a large jacket pocket.

Then, once the presses showed their ability to change the power structure, the fear of this new printing contraption— now churning out more and more new material—began to grow. McCleery says that political and religious leaders panicked over the potential for so many new and varied ideas being shared without their help or approval. One Venetian judge condemned the change with the pronouncement "The pen is a virgin, the printing press a whore."

Although this language is a little colorful for a judge, the fears that spread through society are understandable. In the past, you needed a pen and the ability to write to share your thoughts, opinions, and ideas—even on a limited scale. This changed rapidly when society gained access to a printing press and an individual could reach tens of thousands of literate people. The elite—the clergy and the nobility—had controlled the conversation when they controlled the pen.

The printing press, in comparison, could not be controlled—much as the Internet cannot be controlled today.

This kind of technochondria happens partly because of our fear of the new and in some instances is still prevalent in the power struggles of governments and citizens' freedom. This was visible early in 2010 when a group of Chinese computer hackers managed to breach and steal user information from Google's computer servers in that country. Google believed, from information it obtained, that the hackers were involved with the Chinese government and were trying to gain personal information about individuals who were illegally blogging within China. The Chinese authorities weren't concerned only about the Internet and technology but about what they created: the power of access to unfettered information.

TV Will Rot Your Brain, Don't You Know!

When a development is new and just catching on, we rarely have a clear vision of the future, an understanding of the effects.

We don't really know how to integrate the innovation into our current habits and norms, and we also fear that adopting the new will affect our old ways of doing things. The tension, fear, and anxiety resolve only over long periods of time as we figure out how best to use the new technologies.

Television, for example, was expected to have devastating effects on the printed word and even the arts. A brief 1929 *Washington Post* article reported that meetings were held to discuss whether television would "detract from theatre attendance when it is more fully developed."

Even when these technologies finally break through the obstacles and take off, we don't really know what to do with them. The first television shows, created in the mid-1920s, were essentially filmed radio shows shot with a single camera. They were broadcast initially to the select few lucky homes equipped with the newfangled TVs capable of displaying fuzzy images in black and white. Gradually, the creators moved to three cameras, but they didn't use any dramatic video cuts or special effects. The camera was stationary, and what viewers saw was often nothing more than a radio host, usually sitting behind a desk, puffing on a cigarette, and explaining a story just as if he were talking on the radio.

Early newspaper articles described the television as "radio with pictures," and early TV series were called "radio serials." Some shows were fifteen-minute segments without a single edit or transition. Still, people were mesmerized by television. They didn't need fancy editing to keep their brains occupied. Just the fact that the image was moving was enough to keep a flow of energy zipping around their heads.

It took several decades for the medium to expand, eventually adding drama, comedy, more detailed news, and much fancier camera work—but even that wasn't easy. When fast pacing and multiple cameras created different views on the screen, the old fears reared up again. Hundreds of papers and articles published from the early days of television all the way through to the advent of MTV's fast-paced edits underlined and underscored the fears of parents, politicians, and clergy that television would corrupt and ruin society. Scholars and editorial writers were sure it would destroy our youth, inspire violence and sexual exploitation, and turn our brains to so much oatmeal-like mush. As humans, the reports said, we just weren't built for consuming information this way.

Still, television got off relatively easy, maybe because all the generations enjoyed it. Although it still produces a fair bit of anxiety and concern, the backlash is nothing like the fear-driven fire and brimstone directed at comic books.

Bang! Pow! Bam! Danger Ahead!

Although comic-style illustrations can be traced back more than a thousand years, the genre really started to take form and become a mass medium between the 1920s and 1930s in the United States. Comic books grew dramatically at that time because their creators decided to focus on children, not just adults, and found an audience that could appreciate silly humor and illustrations. As a result, hundreds of new comic-book titles were born in the late 1930s, including the modern superheroes Batman and Superman. Another genre, "gross-out comics," also emerged, with fare that was juvenile in nature and usually centered on crime, especially murder.

These more offensive stories drew the attention of parents and politicians, who came to be convinced that comic books would destroy the young people of the day and drive them to commit horrific crimes—much like the arguments we hear about video games today.

In April 1954, Congress began hearings accusing the comic-book industry of promoting and inciting juvenile delinquency. These hearings, which took place in New York City, were chaired by Robert Hendrickson, a Republican senator from New Jersey and chair of the Senate Subcommittee on Juvenile Delinquency. A Democratic senator from Tennessee, Estes Kefauver, who previously had investigated organized crime, also played a prominent role in the hearings.

In the book *The Ten-Cent Plague: The Great Comic-Book Scare and How It Changed America*, David Hajdu writes that the outcome of the televised and much-publicized hearings was essentially decided before they began. The majority of the "experts" called in to testify were sure that the industry was destroying youth. On the first day of the hearings, Fredric Wertham, a famous psychiatrist known for his expertise on criminals and sex offenders, testified that he was certain, "without any reasonable doubt and without any reservation, that comic books are an important contributing factor in many cases of juvenile delinquency." Wertham even called comics *Superman* and *Tarzan* sadistic and masochistic. Then he went further, saying quietly, "Hitler was a beginner compared to the comic book industry."

After the hearings, at least twelve states developed new anti–comic book laws and oversaw comic-book burnings. Congress urged the industry to police itself, and, feeling the pressure, a new comic-book industry oversight group was created called the Comics Magazines Association of America. This group came up with a set of strict rules, a "Code for Editorial Matter," which make any video-game warnings we have today seem incredibly tame. To protect children of the future, the rules included the following:

Policemen, judges, government officials, and respected institutions shall never be presented in such a way as to create disrespect for established authority.

Criminals shall not be presented so as to be rendered glamorous. . . . In every instance, good shall triumph over evil and the criminal punished for his misdeeds.

Profanity, obscenity, smut, vulgarity, or words or symbols which have acquired undesirable meanings are forbidden.

All characters shall be depicted in dress reasonably acceptable to society.

Ridicule or attack on any religious or racial group is never permissible.

No comic magazine shall use the word "horror" or "terror" in its title.

The treatment of love-romance stories shall emphasize the value of the home and the sanctity of marriage.

All scenes of horror, excessive bloodshed, gory or gruesome crimes, depravity, lust, sadism, masochism shall not be permitted.

All lurid, unsavory, gruesome illustrations shall be
eliminated.

In other words, titles such as "Casper the Friendly Ghost"
were OK, but "Betty Boop" in a halter top or anything with
crime or zombies was not.

Was there any real proof that comics caused juvenile delinquency? No. But fear of and anxiety about something different was enough to blame a burgeoning industry for ill-behaved
children—who, it turns out, had been around long before
comic books were invented.

Computer Shock

Tracing the reaction to the explosion in computing power and
the expansion of the Internet is somewhat like rewinding technochondria's greatest hits. In the short span of a few decades,
we've seen all the familiar fears and skepticism rear up again—
from doubts that computers would produce any benefit to the
belief that the technology will harm or destroy our kids.

In the 1970s, as computers grew smaller and more powerful and terminals began to appear on workers' desks, many
experts still couldn't anticipate the revolution ahead.

Kenneth H. Olsen was an MIT-trained engineer who
founded Digital Equipment Corporation in 1957 and helped
build some of the early successful minicomputers, which allowed individual workers to take advantage of computing
power by using a terminal connected to a midsize computer.

In the early days, Olsen said, "we thought even children could understand computers. We thought computers were fun and we thought they could change the world. But we had no idea that they really would do just that."

Still, even this pioneer and innovator was skeptical about how far the trend would go, telling the magazine *Financial World* in 1976—the same year the very first Apple computer went on sale—that he didn't really see a place for computers in the home. "While a computer might be great and educational for a smart kid, I think we already have too much automation at home," he said. "In general, our lives should be simpler."

Not surprisingly, Digital Equipment mostly missed out on the personal computer boom.

The potential of the Internet generated a similar response. It started out as a way to allow academics and scientists to share information, and back then, it was slow and clunky. But even as it began to bring in all kinds of users, there were those who dismissed its use in the same way the monks had shrugged off the printing press.

In a classic article in *Newsweek* in 1995, Clifford Stoll, an astronomer and author, threw cold water on all the dreamy possibilities that the online world seemed to have: "Visionaries see a future of telecommuting workers, interactive libraries, and multimedia classrooms. They speak of electronic town meetings and virtual communities. Commerce and business will shift from offices and malls to networks and modems."

To this, Stoll had a one-word response: "Baloney."

All those voices online would simply make a lot of noise, he scoffed. And reading or learning online? Preposterous.

"Nicholas Negroponte, director of the MIT Media Lab, predicts that we'll soon buy books and newspapers straight over the Internet. Uh, sure," he wrote.

Just fifteen years ago, he couldn't possibly see how we might buy airline tickets, make restaurant reservations, or negotiate purchases online. And, he added, "Who'd prefer cybersex to the real thing?"

Who indeed?

Stoll was sure that human contact was necessary for sales, communication, and education. Yet now he looks as out of step as those writers who predicted that the telephone and the phonograph would kill the arts and human interaction.

What he missed—and what so many of us have trouble grasping—is how difficult it is to foresee exactly what changes a new technology ultimately will bring. As with the printing press, the biggest changes with computing and the Internet took place when people could take the Web with them rather than having to go somewhere to use it.

Just as pocket-size books brought reading to a greater audience, the handheld BlackBerry brought e-mail to a gadget that easily fit in a person's pocket and made it an indispensable part of daily life. As laptop sales have grown faster than desktop sales and portable machines have gotten cheaper and lighter, the Web has grown exponentially. The Internet, which now has nearly 2 billion users and has grown rapidly, reached only 16.5 million users twenty-five years ago. Similarly, in 1998, as mobile phones began to shrink in size and price, there were only about 4 million active mobile phones in the world. By 2008, when phones were hardly bigger than a pack of gum, that number was 3.8 billion, or almost 70 working mobile phones for every 100 living persons worldwide. In 2009, the number reached 4.6 billion.

The ubiquity of all these gadgets has created a new round of fears and assertions that computers and the Internet are responsible for a raft of societal ills, harming children and adults. For instance, for the better part of the last decade, some teachers and concerned parents have argued that Internet WiFi is damaging to our health, even calling the output from electronics and WiFi "electrosmog." In 2008, some schools and offices banned all forms of wireless Internet—even though there isn't a shred of evidence that WiFi specifically is responsible for any health issues. Lakehead University in Canada, which adopted a ban, proclaimed that WiFi could cause "potential chronic exposure for our students" from electromagnetic rays and asserted that the risks of WiFi are equal to those of secondhand tobacco smoke. Yet studies show that older technologies such as televisions, microwaves, and radios emit stronger electronic waves than do WiFi hubs.

There are also concerns about the ergonomic effects of computers, alarm about the corrupting impact of Google, and worries that the next generation of computer-addicted children will be unable to navigate society properly.

A wave of books have argued that computing, the Internet, and screens are going to bring with them the demise of society and a youth so corrupted that it will only be able to watch MTV and look at picture books. In the mid-1990s, in the book *The Gutenberg Elegies: The Fate of Reading in an Electronic Age*, Sven Birkerts questioned whether the digital age would produce illiterate children who are unable to read long-form literary works and are capable only of passively watching images on screens.

Maggie Jackson, in *Distracted: The Erosion of Attention and the Coming Dark Age*, argues that multitasking is so bad for society that it could put us back in the Dark Ages, unable to interact with one another and incapable of experiencing meaningful and intimate relationships.

Lee Siegel, a cultural critic, in *Against the Machine: Being Human in the Age of the Electronic Mob*, suggests that heavy Internet users are destined for a life of technological solitude so bleak that our humanity and individuality could dissipate into the ether.

The members of a group called the Alliance for Childhood are well-respected psychiatrists and childhood development professors who regularly release reports alleging that computers are ruining our youth. The group's mission statement proclaims, "The lure of electronic entertainment diminishes active play and work and the learning of hands-on skills," and when it comes to technology and children, "the losses often outweigh the gains." An older report written by the group titled "Fool's Gold: A Critical Look at Computers in Childhood" concludes: "Computers pose serious health hazards to children. The risks include repetitive stress injuries, eyestrain, obesity, social isolation, and, for some, long-term physical, emotional, or intellectual developmental damage."

As should be clear by now, such worries are part of the territory. In all fairness, sometimes they are legitimate: The printing press did shift power away from the clergy and monarchs, and the Internet is giving voice to a broader range of people, kooks and creeps included. It's perfectly normal and probably healthy to examine whether these changes are good or bad. But we'll also no doubt look back at many of the debates a generation from now and see that a lot of these fears were inflated and maybe a bit ridiculous, too.

Long Form Has a Long Life Ahead (31 Characters)

When we adopt a new way of doing something, we also have to give up the old comfortable ways we're accustomed to, and that kind of change frequently comes with its own anxiety.

In recent years, an increasing amount of information seems to be streaming into the world byte by byte—in text messages on your phone, tweets and status updates from your friends, and headlines swimming across your television screen and on your Google home page. By early 2010, 50 million "tweets" were moving each day via Twitter, the social networking site where people can send messages of up to 140 characters at a time to "followers." More than 700 million times a week, friends shared shortened links of videos, stories, and websites. The sheer volume of staccato messages, coupled with the sheer volume of information coming at us from a gazillion different directions, has created yet another worry: Will long-form content—the snacks and meals of an educated society—die, leaving a culture that can only graze in small byte-size pieces?

No. Absolutely not.

As we've seen, throughout history we have tended to dramatize the death of one form of communication when another is being born.

Sure, there is clearly an abundance of short-form material, but let's be realistic: This isn't the first time we've communicated in a few words. Newspaper headlines have never offered a lot of verbiage. Radio stories and television stories are surprisingly brief when written out. And honestly, when was the last time you abandoned a book because the table of contents sated your thirst for knowledge?

Maybe you don't read as many books or watch as many hour-long television shows as you used to because you are doing other things, such as playing video games and catching up with DVDs or downloads on your computer.

Given all the hue and cry, it's important to take a different look at history. Before there were even screens in our living rooms, the same worries reared their heads. There was a time in the 1920s when cultural critics feared Americans were losing their ability to swallow a long, thoughtful novel or even a detailed magazine piece.

The evil culprit: *Reader's Digest.*

Before Twitter, There Was Dewitt Wallace

In 2009 I gave a talk titled "The Future of News" at several conferences across the country. The presentation usually lasted twenty minutes and covered most of the innovative work happening inside the *New York Times* as well as other technological innovations in journalism. I tried to assure the conference attendees that long-form journalism might look different from how it does today, but it would survive well into the future.

Without fail, at the end of each talk someone would cite Twitter or another short-form technology as the signal that the death of the long form was upon us. At an event in Boston, one attendee argued that "one day, there won't be any more long-form books or news articles—instead everything will be the length of the *Reader's Digest*."

I offered the number of investigative journalism books on the bestseller lists and the high number of page views for long-form articles on the *Times* website as a rebuttal, but the question got me thinking. Could *Reader's Digest* really be the model for our future? That question led me to DeWitt Wallace.

Early in the twentieth century, while recuperating from a World War I injury, young DeWitt Wallace was confined to a hospital bed in France for more than four months. He had little else to do but read stacks of magazines from America. Toward the end of his hospital stay, he concluded, rightly, that most people were too busy to read all the wonderful material printed each month. But he came up with a solution: He could condense the best articles and reprint them together in a special "reader's digest."

On his return to the United States, Wallace put together a business plan for a magazine that would condense the best articles from American magazines. The notable publishing and business tycoons of the day dismissed his idea as "too niche," and bankers refused to fund it, saying *Reader's Digest* couldn't possibly gain an audience of more than 300,000 readers.

But Wallace was confident and passionate and didn't give up easily. He found a business partner—who later became his wife—and in February 1922, his magazine went to press. The first issue of *Reader's Digest* contained thirty-one articles, one for each day of the month, all selected and edited by Wallace and condensed into one or two pages in his pocket-size reader.

By 1929, circulation had grown steadily to 200,000. Then it exploded, expanding to just under 1.5 million by 1935, a sevenfold increase in five years. Had Wallace found a magic potion? Was the public concluding that condensed content—reading lite—was the future?

Not exactly.

Sure, the articles were shortish, in big type, and on small pages, so they felt easy to read. But the length of the articles wasn't the attraction. James Playsted Wood, who wrote a history of *Reader's Digest,* noted that "above all, the magazine accentuates the positive, minimizes the negative and strikes a note of hope whenever possible." People weren't buying Wallace's magazine for its shorter stories. Rather, they wanted a homogeneous, religiously and politically conservative magazine. And that's what they got, with stories such as "Whatever Is New for Women Is Wrong," "What People Laugh At," and "Is the Stage Too Vulgar?"

The digest was criticized for its story length too after a reviewer pointed out that some of the "condensed" versions of stories were actually longer than the original magazine articles. Critics also believed that the articles weren't chosen for their literary or journalistic merit but on the basis of their simplicity and whether the point of view aligned with Wallace's conservative bent.

By the 1930s, some magazine publishers were so annoyed with *Reader's Digest* that they threatened to block Wallace from condensing their articles. They believed the magazine wasn't just a low-calorie version of their content but a rewriting of work that didn't match Wallace's views.

In response, DeWitt Wallace decided to hire his own writers to create his own stories—but the practice evolved in an unusual way. At first Wallace hired a few writers and started to assign and create original stories for the magazine. But he soon realized that he was changing the nature of his publication. So instead, he began to "plant" longer stories in other publications, paying other magazines to run his writers' work. *Reader's Digest* could then excerpt those planted articles. Wallace came to understand that he was providing "first-person editing," stories plucked and polished to satisfy his particular audience, just as today's explosion of content allows you to develop your own personalized newsreel.

In 1945, the *New Yorker* published a five-part investigative series about the *Reader's Digest's* strange "planting" practices and noted that the magazine had, over the course of six years, run 720 condensed reprints from other publications, 316 articles written solely for its publication and described as such, and 682 stories written specifically so that they could be excerpted in the digest. In other words, nearly 1,000 articles had been assigned and written at Wallace's direction. The *New Yorker* discovered that more than sixty publications had been paid to run articles so that they could be reconstituted in *Reader's Digest*. Later, it was discovered that through the 1940s and 1950s, three out of five articles in *Reader's Digest* were actually original content commissioned and edited by Wallace.

Although it wasn't necessarily clear sixty years ago, it's clear now that the magazine's popularity reflected its subject matter. Readers weren't abandoning long stories for short ones; they were gravitating to Wallace's touch with happy-go-lucky, sanguine, politically conservative articles, well packaged in a little pocket reader. In short, just as with pornographic movies, the appeal of *Reader's Digest* was in the overall experience.

Even now, *Reader's Digest*, with its byte-sized approach, has a circulation of about 8 million. The full-meal *New Yorker*, by contrast, has 1 million.

Irtnog

There was another lesson that grew out of the era's fears that Americans would ditch their novels and thoughtful magazine pieces for the slick, short fare of the *Reader's Digest*: In the rush to adopt new ideas and innovations, we sometimes go overboard, driven not so much by the joy of discovery as by the nagging fear that maybe we will miss something important.

In a clever satirical piece in the *New Yorker* in 1938, the influential essayist E. B. White captured this classic human response. White told the make-believe story of readers who were so determined to keep up with the exploding number of magazines and newspapers that they sound like people e-mailing and texting on their way to work. They "read while shaving in the morning and while waiting for trains and while riding on trains. . . . Motormen of trolley cars read while they waited on the switch. Errand boys read while walking from the corner of Thirty-ninth and Madison to the corner of Twenty-fifth and Broadway."

The *Reader's Digest* offered an alternative, White wrote, and others started digests too, hoping to capture the original's success. "By 1939 there were one hundred and seventy-three digests, or short cuts, in America, and even if a man read nothing but digests of selected material, and read continuously, he couldn't keep up," White wrote, continuing the gag. "It was obvious that something more concentrated than digests would have to come along to take up the slack.

"It did. Someone conceived the idea of digesting the digests. He brought out a little publication called *Pith*, no bigger than your thumb."

Still, that wasn't enough. So "*Distillate* came along, a superdigest which condensed a Hemingway novel to the single word 'Bang!' and reduced a long article about the problem of the unruly child to the words, 'Hit him.' "

Ultimately, White went on, a graduate student figured out how to condense everything into a six-letter word: "Everything that had been written during the first day of his formula came down to the word 'Irtnog.' The second day, everything reduced to 'Efsitz.' People accepted these mathematical distillations; and strangely enough, or perhaps not strangely at all, people were thoroughly satisfied—which would lead one to believe that what readers really craved was not so much the contents of books, magazines, and papers as the assurance that they were not missing anything."

Not missing anything? Ultimately, the digests played on this unavoidable tension between the new and the old: If you don't get on board—big time—you will be out of the loop or left behind. Linda Stone, a prominent technologist who spent nearly two decades as an executive at Apple and Microsoft, sees the same worry today. When you compulsively check e-mail, or run to the mailbox, or open up Facebook, she says, you aren't simply being obsessive or trying to avoid work. You've succumbed to something much deeper. Stone calls this "continuous partial attention": a need to know what's coming next, an "effort not to miss anything."

So it has been with Facebook, a service originally intended for students that in recent years has added millions of middle-aged users afraid of missing out on a technological phenomenon. When Twitter came along, the same fear prompted millions more to jump in and tweet 140 characters about just about anything, though many of them weren't sure why they were doing it. Whether these will be lasting, meaningful innovations or ephemeral trends isn't yet clear. Given the parallels, though, I'd be tempted to predict that the next big thing will be our own made-up language, our own Irtnog and Efsitz.

But then, I realized, we already have that, too.

Text Me

Over and over, I read in newspapers and research reports and hear on television, at conferences, and around the dinner table that our language is deteriorating. People proclaim that kids don't use proper English anymore, that they communicate only in a broken, acronym-style speech. Some believe that the members of the next generation are destined to be at a disadvantage when they have to work with or compete against those who can write "correct" English.

A quick search of the Web will conjure up thousands of articles about the death of our language. In 2008, for example, the British newspaper *The Guardian* complained about the overuse of the exclamation mark and LOLspeak. The result, it suggested, would be that people eventually will write "whole emails using these things, communicating like two fax machines and rendering words obsolete."

Wired, the technology magazine, in 2005 pointed to a series of studies about the use of these acronyms, noting that "traditional linguists fear the Internet damages our ability to articulate properly." Although *Wired* didn't see the future as negatively as most, it clearly was highlighting questions about the future of language.

Behind the worries is a strange assumption that language is fixed and unchanging, that all these funky abbreviated words result uniquely from the byte-sized-communicating, social-networking, video-game-playing, iPhone-toting ways of the Internet Age. But the acronym isn't a product of the digital generation. Acronyms, abbreviations, and shortcuts have been a part of language since . . . well, ever since language has existed.

Some references date back hundreds of years, such as B.C. for Before Christ and A.D., anno domini, for in the year of our Lord. Medical and military professionals are especially acronym-obsessed, giving us HIV, IQ, DNA, Humvee, SWAT, and POW. Of course, acronyms have also come to us through technology, with words like "radar," along with constructions such as "VHS" and "hi-fi"—all condensed versions of a longer string of words.

Many other words we use every day that are now accepted as correct were much longer in the past. The popular word "pub" comes from "public house." A "bus" was once an "omnibus." Scuba diving comes from the long technical term "self-contained underwater breathing apparatus," and clearly, the abbreviated version rolls off the tongue more easily, especially under water.

If this is all old news, then OMG, why are so many people worrying about gr8 and LOL and IMHO in this latest incarnation of acronyms?

One reason may be that the changes happened with unusual speed. But another may be that this new communication is fundamentally different from anything we've known in the past.

Most linguists agree that language serves two purposes. One is to write, to record history on paper, share ideas, or make note of events. Writing's main function, well beyond grocery lists and phone messages, is to record more complex stories and their details.

By contrast, we speak predominantly for dialogue, to exchange information with one another. Technology hasn't really changed this use of language since we first started talking to one another in caves many, many thousands of years ago. The telephone didn't change that either; dialogue still had to take place with speech.

But now, with instant messenger applications, text messaging on mobile phones, and instantaneous e-mails, the Internet has torn down the distinctions between speech and writing. For the first time, society as a whole has engaged in real-time conversations using text, merging writing with speech. That has created something of a new language.

New acronyms help us bridge the differences between the written word and the spoken one. For example, if you are chatting with a friend online and she tells you a joke, you need to let her know that you got the punch line. To solve this problem, people started using the acronym LOL to explain that they were "laughing out loud."

If you walk away from your computer in the middle of a chat, the person on the other end won't understand your silence. Somewhere along the line someone typed "BRB" into a message window to alert the other person that he would "be right back." Without that polite explanation, the screen goes eerily quiet and the recipient feels dismissed.

Although many acronyms don't graduate from individualized banter among friends to widespread use, there are many, many new acronyms and language adjustments morphing and melding all the time through our digital gateways. Some catch on and become de facto standards, such as LOL and BRB, and some wither away or stay confined to small groups. Take the acronym ASL, for example. In the early days of the Web, those letters were used to ask a person's "Age, Sex, and Location" on an instant messenger client. Now most social networks require people to pick an image for their icon and this question is answered by glancing at a person's photo.

David Crystal, a linguist and writer on the new "text speech" or "netspeak," doesn't believe the abbreviations such as R for "are" and symbols such as :-/ for "indifference" are causing language to deteriorate. Rather, he sees them merely as a function of current technology's limits, and a temporary one at that. "The whole point of the style is to suit a particular technology where space is at a premium, and when that constraint is dropped, abbreviated language no longer has any purpose," he writes.

Jesse Sheidlower, editor at large (North America) of the

Oxford English Dictionary, also sees the new words simply as a natural progression of language in society. These language adjustments happen all the time, Sheidlower said to me in an interview, "People always have vocabulary differences. Every generation creates words they develop and use for different occasions. Some will live on and some will die out, but it's just a natural progression of our language." Sheidlower pointed to the word "OK," which today can be used in any number of settings, and although there are numerous theories of the word's origin, some believe it points to the words "ol korrect" which today would mean "all correct."

Sheidlower doesn't see acronyms or new words changing our current forms of conversation, saying, "I don't think this is going to affect our language as such, but it does offer a different way to communicate, and in general I think that the more ways one has to communicate, the better."

These changes, he explained, will always happen from the bottom up in society, not from the top down. When he adds a new word to the *Oxford English Dictionary*, the word comes from daily use in verbal and written communication, not from scholars sitting around a table. Take the word "crunk." Recently added to the *Oxford English Dictionary*, it means "a type of hip-hop or rap music characterized by repeated shouted catchphrases and elements typical of electronic dance music, such as prominent bass." It's pretty apparent from its meaning that this word wasn't invented by ivory-tower academics but bubbled up from the bottom, from the vernacular of the day.

Even the verb "Google," which means "to use the Google search engine to obtain information about (as a person) on the

World Wide Web," became an entry the *Merriam-Webster Dictionary* in 2006. This didn't happen because the search giant petitioned for a new word but because the word was being used so often in that way that it simply became a de facto part of the language. (The same year "Google" became a verb, the words "biodiesel," "spyware," "hacktivism," "uninstall," "texting," and "ringtone" were added to either the *Merriam-Webster* or the *Oxford English Dictionary*.)

Even as young people develop their own words, research shows that they understand how to converse with different audiences. In a recent study, students at the University of Colorado chatted on an instant messenger with their friends and then with school librarians. Not surprisingly, the conversations with the librarians were more formal than those with other students and friends, though they all took place over an instant messenger.

Rather than lament the use of acronyms on mobile phones, in e-mail, and through instant messenger applications, the world should acknowledge that these kids are helping develop a new type of cultural communication. These byte-size consumers at the bottom of the language food chain are helping create a vernacular that can be shared equally and equitably by an entire community of texters, video chatters, micromessagers, and e-mailers of all ages.

You can lament the changes that are happening today—tomorrow's history—convincing yourselves of the negatives and refusing to be a part of a constantly changing culture. Or you can shake off your technochondria and embrace and accept that the positive metamorphosis will continue to happen, as it has so many times before. Young people today are building a new language, not demolishing an old one. And as you will soon see, developments like these new words are helping create significant and meaningful new communities and new relationships that are an essential part of our changing culture and our wireless future.

3

your cognitive road map
anchoring communities

I had a *doppelgänger* living in the same area of Brooklyn.

Meet Sam H., My Good Friend. Sort of.

I've never met my friend Sam H. I don't know what he looks like and wouldn't recognize him if we were in the same room. Still, I can assure you that Sam H. is real. In fact, I consider him a good friend, though I don't even know his last name.

We met mostly by chance. As a gadget and game lover with a sophisticated smart phone, I enjoy playing an online game called Foursquare, which involves marking my location whenever I arrive at a store, restaurant, or park. The old-fashioned four-square game was often played with four kids at a time, a ball, and a court with four squares, usually drawn with chalk on a playground or residential street. The online version of this game is also interactive, but it requires a Web connection and can involve a lot more than four players. It's a cross between a location-based game and a "where-are-my-friends" experience.

The interactive version of Foursquare was created in 2009 by Dennis Crowley and Naveen Selvadurai, two New York–based computer programmers. Crowley, an energetic shaggy-haired programmer and entrepreneur in his mid-thirties, is the CEO of the company and has spent the last ten years working on interactive location-based games of one form or another. As with most start-ups, Foursquare is the result of inquisition and serendipity. While planning a trip to Scandanavia in 2008, he grew frustrated after a Google search netted results that were random and therefore not very useful. He then reached out to his friends to ask for travel tips and recommendations and posted a quick question on the social photo-sharing site Flickr.com, asking if people could suggest interesting places to visit in Scandinavia. "I got tons of amazing responses," Crowley said. "People said, check out this museum near the shipyards, or go to this coffee shop and check out the amazing statues in the basement, and if you go to these shipyards, make sure you stand at a specific angle and you'll see an old shipwreck." The result: a magical trip that wouldn't have been so if he had relied solely on the details spit out from a Web search.

Crowley decided to build an application that would allow users to share fun facts about locations and add gamelike elements to the experience.

The resulting Foursquare is one of a growing number of

mobile phone applications that appeared in 2009 to take advantage of a smart phone's ability to pinpoint a person's precise location. For instance, real-estate applications can help buyers find homes where they are now. (Imagine walking around a neighborhood and using your phone to explore homes for sale around you.) Google offers a service called Google Latitude that allows people to share their locations with friends. Twitter lets users append their location to tweets. Other applications can be used to gather information about the community around you, such as its schools, its medical services, and even the best coffee shop. In providing information in a location-specific way, these apps also allow companies to deliver highly specific advertising or even coupons directly to a person's mobile phone.

To play Foursquare, I start up its application on my phone whenever I arrive at a restaurant, bar, café, or park and "check in." My check-in tells my Foursquare friends where I am at the moment and gives me points for my good (or bad) taste. I can add reviews or recommend the daily specials. But the real fun is the game part: I earn Boy Scout–like badges for multiple check-ins in a day, for visiting the park with my dog, for stopping at the karaoke bar, and so on. Even better, if I'm the most frequent visitor at a certain store or restaurant, Foursquare names me the "mayor" of that place. Mayors usually don't get anything tangible for their regular stops (though some restaurants, such as Starbucks, offer discounts or freebies to the local mayor), but they always get bragging rights. That in itself can be a strong incentive to keep checking in everywhere you go.

Which brings me back to my friend Sam H.

Near my home in Brooklyn there's a coffee shop called

Southside. I sometimes go there several times a day to satisfy my coffee addiction, and each time, I check in on Foursquare. With more than sixty check-ins in a month, I had the unique and proud distinction of being Southside's mayor. Until recently.

One morning, I walked to the coffee shop, ordered my coffee, and pulled out my phone to check in. But instead of being greeted by the usual alert that announced, "Congratulations, Nick, you're still the mayor of Southside!" I got a shocking new message: "Thanks for the check-in. Sam H. is now the mayor of Southside!"

I immediately assumed something was wrong with the Foursquare database. After all, I had been checking in at Southside more than once a day for months. The software had to have a bug in it.

I finished my coffee a little more quickly than normal, rushed home, opened my laptop, and looked up Sam H. on the Web. To my surprise and dismay, I learned not only that he had been crowned the new mayor of Southside but also that we frequented many of the same bars, restaurants, and coffee shops. A bit more searching turned up that he teaches at New York University, just as I do.

I had a doppelgänger living in the same area of Brooklyn.

Being competitive and proud of my mayorship, I couldn't leave well enough alone. Though I was probably breaking the unwritten rules of the game, I found Sam's e-mail address and sent him a note, demanding (in a friendly way) to know why he had stolen my mayorship.

In the same spirit, he wrote back, telling me to keep away from his neighborhood, that he was the new mayor in town. We went back and forth like this for weeks, stealing each other's mayorships and "arguing" like an old married couple fighting over the remote control.

And we became friends, at least online. We connected on other social networks, such as Facebook, Twitter, and Flickr, and we regularly communicate back and forth about new restaurants, bars, and other hot spots in our shared neighborhood.

The funny thing is that I've only seen a few small and blurry photos of Sam H. on Facebook and Twitter. I surely couldn't pick him out of a crowd or even identify him in a lineup. But we share experiences and communicate as much as I do with friends at work.

If you and I went through my address book, I could share many stories like this. Maria is a very good friend I met online a couple of years ago who lives in Bulgaria. Since our meeting on some social file-sharing sites, we've actually met in person twice, each time for less than an hour. But I don't question the authenticity of our friendship just because our encounters are digital; instead, I appreciate the camaraderie as well as the interesting, though geeky, technology and media stories we exchange online. Jason is a whiz at finding fun and interesting design news. Though we met only once at a conference, he lives in San Francisco, and I might not be able to pull him out of a police lineup, I trust his design-news judgment more than that of some of my colleagues at the *Times* and NYU.

I don't see any lines between real-life friendships that involve talking or looking someone in the eye and virtual ones in which the communication is through e-mail or text messages. Any of those relationships can be good friendships. We may not drink beer or coffee together or exchange birthday and anniversary cards. But we can send pictures and admire each other's pets via photo albums on Facebook, send birthday greetings, or share funny videos and important news via Twitter. One experience doesn't supplant the other; instead, together they create new bonds and friendships we might not have experienced otherwise.

Because of these relationships, those somewhat unknown online friends may be as influential—or more so—as a running buddy or a next-door neighbor. You and I are just as likely to accept their recommendations for restaurants and plumbers. They may influence the books you read, the movies you see, or the news you click to. Because you know they have common interests, you may trust them even if you don't know them well enough to describe their hair color or favorite sports teams. As a result, these new communities and their members have a powerful and growing impact on what businesses their "friends" frequent, what they do, and how they spend their money. In the future, their power is going to grow in expected and unexpected ways.

Already these relationships have become filters for the content that appears on my digital doorstep. Take my Sunday reading experience: Several years ago my wife and I would lie in bed on a Sunday morning with the *New York Times* print newspaper and a few weekly magazines. Now, every night before we go to sleep and every morning when we wake up, we browse our mobile phones or laptops, looking at the information our communities share with us, and in turn share interesting information with them. These links bubble up from our personal connections on broad networks rather than being imposed by a faceless curator. Instead of our relying on professional editors to package a home page or produce a printed page, our on-line friends are now our de facto editors, providing a supply of news and information that is highly personalized and tailored to our interests. As a result, these relationships are much more than "social." They are hugely influential.

Defining Communities

As one of the Web's most popular buzzwords, a "social net-work" most often means a website or service that enables people to communicate or be in touch with one another in a personal way. Facebook, for example, is one of the largest so-cial networks, with hundreds of millions of users.

When the term "social networks" started gaining traction online, it seemed redundant to me. The Web was supposed to spur social exchange—that's why it was created, so that people could communicate and share information with others. Plus, many early Web users, me included, had been chatting and sharing images and content on message boards, on forums, and in other dark alleys of the Web for years.

As the label "social" started to spread into job listings, ré-sumés, and advertisements, I kept thinking there was more to the idea of people being social on a network. I mean, if you built homes on a new street and strangers moved in, would you be shocked when they all started to talk to one another? When the people living on the street started having dinner parties and talking about books they found interesting or movies they had seen, would you hire anthropologists and scientists to interview everyone? Probably not. In fact, we would probably be shocked if they *didn't* start communicating and being "social" with one another.

That doesn't mean that I don't think social networks are important. Quite the contrary. As you can see with Sam H., I think these networks have a far more significant role than just being a connection between people or a way to tell the world what I had for breakfast or even share links. But it wasn't until I read *Imagined Communities: Reflections on the Origin and Spread of Nationalism* by Benedict Anderson, a Cornell University emeritus professor of government, that I gained a real understanding of what was happening online with our social networks.

Anderson has spent most of his career exploring, breaking down, and defining what it means to be a nation. His work has been incredibly influential in creating a new explanation of nationalism and the building of nations. As I learned about his theories, it occurred to me that they also unintentionally applied to the Internet, which, in a way, is a nation all its own.

In the 1980s, Anderson went way beyond conventional terminology and developed a fascinating and groundbreaking theory, proposing a new definition of a nation. "It is an imagined political community—and imagined as both inherently limited and sovereign," he wrote. "It is imagined because the members of even the smallest nation will never know most of their fellow-members, meet them, or even hear of them, yet in the minds of each lives the image of their communion."

In your life, you have all kinds of these communities. The nation you live in is one, of course—your passport proves this. But so are your church, your neighborhood, and your alma mater. Anderson would argue that actual communities exist only when the other members of the community are physically present for us, there in the flesh to be seen, as in church on Sunday morning or Yankee Stadium in the summertime. But faced with an inability to be physically aware of all the others in a community, we imagine their existence.

Our physical locations help illustrate this thinking a little better. Though I will never meet or know even a small fraction of the people who live in America, I feel connected by our shared belief in our Americanness. I feel a strong sense of camaraderie and esprit de corps with the more than 300 million people who have the same passport I do, but that sense of community exists only in my imagination, as it may in the imaginations of my fellow citizens.

New York City is another imagined community. Brooklyn, the borough of New York City where I live, and Thirty-third Street, where my house is, are imagined communities as well. Even if I devoted my entire life to trying to meet everyone in the community of New York City, it wouldn't be humanly possible. I would have to interact with more than 400 people a day for seventy-five years. Yet I still consider all the inhabitants of this city a part of my world, and they consider me a part of their world.

Much of Anderson's work on nations as imagined communities owes its origins to the printing press, which, he says, made the idea of the modern nation possible to begin with. That's the case because the press made books available in the common languages of the men and women of Europe—English, French, and Spanish—instead of Latin. After that, books in a common language became a vehicle for helping a community define its shared ambitions, and the modern nations we know today developed.

In addition, the concept of imagined communities goes way beyond geography: I am middle class, a meat eater, a rock climber, an NYU instructor, a reader of a certain genre of books, a drinker of a specific brand of coffee, and a devotee of the *New York Times*. These all represent different, but important, imagined communities for me. Some are connected and overlap, but most don't, and they are all dynamic, subject to influence by other communities in my life.

Anderson's thesis applies to our online digital lives too. As technology continues to expand and strengthen personal, professional, and social connections across space and time, the ties you feel to your online communities—to people like Sam H.—will grow as well.

At the heart of Anderson's idea is the question of which people we identify with and why. Isn't it possible that I have more in common with a rock climber from China than with a non-rock-climbing American? Could my daily reading of the *New York Times* include real and imagined connections with "like-minded" readers of the same and similar publications? Even reading this book, you're entering an imagined community with others who have read it or will read it in the future— but you'll never know all of them. And although we don't think about it consciously, every story we engage with has some sort of community aspect to it.

Anderson singles out the newspaper for special attention, examining randomly a sample front page of the *New York Times*. The stories, he notes, are diverse: A single front page might have stories "about the Soviet dissidents, famine in

Mali, a gruesome murder, a coup in Iraq, the discovery of a rare fossil in Zimbabwe, and a speech by [then French president] Mitterrand." So what connects these things? he asks, and then answers:

"Not sheer caprice. Yet obviously most of them happen independently, without the actors being aware of each other or of what the others are up to. The arbitrariness of their inclusion and the juxtaposition shows that the linkage between them is imagined."

Sure, he explains, one key connection is the date—all these stories happened or came to light at this single point in time. But all those things were also important and newsworthy, making each paper a kind of "one-day bestseller," with wide influence. And then there was also a shared community of readers.

"Each communicant is well aware the ceremony he performs is being replicated by thousands (or millions) of others of whose existence he is confident, yet of whose identity he has not the slightest notion," Anderson says.

If you and I read the *New York Times*, we are joined together by the information simultaneously presented to us. The newspaper is a community built partially on political interests and opinion but also around the collection of stories, their event date, and the location of the establishment that produces them.

The same could be said for the Bits blog I write for the *Times*. It may touch on several unrelated topics on a particular day, but it speaks to a specific community of readers who may be in the technology business, investing in technology, or just fascinated with gadgets and innovation. Without subscriptions or passwords, defining the exact community that reads the blog isn't easy. A decade ago that community may have existed in smaller slices—say, in subscribers to the *Times* or a technology trade magazine or some arcane Internet bulletin board—but now they can join in one place and, in a completely new way, actually talk to me and one another through our comments sections.

Perhaps the most dramatic example of this new kind of community emerged when Michael Jackson died suddenly and unexpectedly in mid-2009. The tidal wave of response was enormous. According to CNN.com, in an article aptly titled "Jackson Dies, Almost Takes Internet with Him," the websites of TMZ and the *Los Angeles Times*, which broke various parts of the story, both crashed. Google News users couldn't access the news for a period of time. For several hours, Google's top 100 search terms were almost all related to Jackson. Google's "trend" service rated the response to the story as "volcanic."

Keynote Systems, which tracks how websites perform, said that major news sites took more than twice as long to download news stories. CNN said its site had 20 million page views in the hour after the news broke. Wikipedia reported more than five hundred edits to Jackson's entry in the day after he passed away.

In a different generation, small groups might have gathered around a television or radio and maybe later attended a memorial service or sent a letter (with a stamp!). On this day, AOL's Instant Messenger service went down for forty minutes as people tried to communicate with their own networks. Twitter messages topped 200,000 an hour.

In the minutes and hours after the sad Jackson news broke,

enormous, invisible, imaginary, and yet very clear communities formed. Regina Lewis, AOL's consumer adviser, said individuals had three reactions: They wanted to know the news, they wanted to share it, and they wanted to react with their own tributes and remembrances.

On that day, people around the world were connected in unforeseen ways with people they never imagined a connection with. At any given moment, you can feel intimately connected and yet be unable to grasp the other participants in these online groups. Online, communities seem to exist everywhere, behind every website, social network, e-mail address, or news article. In the digital, always-on, real-time, creating, consuming society we live in today, we are constantly weaving in and out of small and large, obvious and imagined communities.

In the same way that Anderson recognized that the printing press and its ability to communicate in a common person's language could break up power structures and create meaningful and powerful nations, so too may our online communities reshape and remake both our own personal imagined nations and our traditional ways of communicating. As the printing press took off, it rightfully frightened those in control, creating fear and anxiety about how society might turn out if so many other people were well informed. Similarly, these extensive new communities and their odd ways of communicating in bytes and snacks, tweets and links, have unsettled those who fear that this will transform our broader nations into a teeming Tower of Babel with lots of voices and noise but little deep thinking. That brings me to another person I recently met online: George Packer.

Bilton Versus Packer, a Twitter Tussle

As the Michael Jackson experience underscored, the Web offers a continual and gargantuan influx of creation and information, and it continues to grow at tremendous rates every day, leading to a natural feeling of information overload.

Consider what happens on Facebook, for example. On any given month, each user creates an average of seventy pieces of content. Altogether for the site's half a billion users, that's close to 35 billion links, news stories, random blog posts, and pictures and videos of friends and loved ones. YouTube, the popular video site, said in 2010 that every single minute, twenty-four hours of video is uploaded to YouTube's servers. That means in a single day 34,560 hours are added to the site—so much that it would take you nearly four years of nonstop viewing to watch all of it.

It's enough to make you want to crawl under a blanket and curl up with a good old-fashioned book.

At least that seems to be pretty much how George Packer felt in early 2010. Packer has covered the war in Iraq, atrocities in Sierra Leone, and unrest in the Ivory Coast; has written several novels and books, including *The Assassin's Gate: America in Iraq*; and has been a *New Yorker* staff writer since 2003. In the face of this avalanche of stuff, he vented his frustration in a blog post.

"Every time I hear about Twitter I want to yell Stop. The notion of sending and getting brief updates to and from dozens or thousands of people every few minutes is an image from information hell," he wrote. "I'm told Twitter is a river into which I can dip my cup whenever I want. But that supposes we're all kneeling on the banks. In fact, if you're at all like me, you're trying to keep your footing out in midstream, with the water level always dangerously close to your nostrils. Twitter sounds less like sipping than drowning."

He wrote that he was particularly concerned by a *New York Times* column by media critic David Carr, my colleague at the *Times*. Carr had written, "There is always something more interesting on Twitter than whatever you happen to be working on."

Well, of course, Packer wrote: "Who doesn't want to be taken out of the boredom or sameness or pain of the present at any given moment? That's what drugs are for, and that's why people become addicted to them. . . . Twitter is crack for media addicts. It scares me, not because I'm morally superior to it, but because I don't think I could handle it. I'm afraid I'd end up letting my son go hungry."

He went on to confess that he doesn't own a BlackBerry or a smart phone and that when he takes the train from New York to Washington, he sits in the quiet car without his laptop and telephone and hopes he can still muster the attention to read for two hours.

His plea struck a chord. And in an ironic twist, a link to his essay was retweeted more than seven hundred times on Twitter.

I can completely understand why Packer is pushing back against a flood of online information. As he notes, there are only a certain number of hours in the day, and given the choice, he'd rather spend those hours on something other than other people's 140-character musings. He's not alone: In comments on his article, many heartily agreed with him.

But after finding his article on Twitter, I wrote a blog post suggesting that he ought to at least give Twitter a tweet. In profound and unexpected ways, the service has the potential to transform news and communications. For example, when Iranians took to the streets to protest their presidential election in summer 2009, the main television news networks reported only sporadically on the response. But people in Iran, who weren't always able to send e-mail messages or upload videos or even access the Internet from a computer, found they could send tweets from their phones. They began sharing details of the emotional response—citizens breaking into stores, people starting fires, and the resulting police beatings—in as much detail as 140 characters could convey. Viewers, reading the dramatic accounts headline by headline and sensing that a rebellion on the order of Tiananmen Square might be under way, complained clearly and loudly to CNN and others.

Using the label #CNNfail, tens of thousands of people vented about the lousy coverage. Some noted that CNN had shown a repeat of Larry King's interviews with the folks on *American Chopper*, the reality television show about people who build motorcycles. Others complained that a spat between Sarah Palin and David Letterman over a poorly worded joke on a late-night show was getting far more attention. Twitter both underlined and highlighted what viewers were missing and gave them a forum for making their intense displeasure known.

Perhaps going a bit too far, I recalled how some journalists feared the potential destruction caused by the railroads and suggested that if Packer had been around 150 years ago, he might have been "afraid to engage in an evolving society and demanding that the trains be stopped."

Clearly, I was stepping into tender territory. My blog post generated more than a hundred comments. As I expected, readers favored Packer's view about 80 percent of the time. One commenter said, "I firmly agree with Mr. Packer! Twitter Dee Twitter DUMB!" Another asked me to "please spare us the implication that we (non-reporters really) need or must have this latest, up-to-date information in order to function in life. That notion is complete hogwash and goes hand-in-hand with the myth of progress as inevitably leading us to the ultimate manifestation of human greatness. I'll take a great investigative journalist or writer (like George Packer) over 10,000 tweets any day."

Packer wasn't any more convinced than the readers despite my arguments about Iran or how Twitter had helped connect families after the earthquake in Haiti. In a follow-up essay, he worried that "[t]here's no way for readers to be online, surfing, e-mailing, posting, tweeting, reading tweets, and soon enough doing the thing that will come after Twitter, without paying a high price in available time, attention span, reading comprehension, and experience of the immediately surrounding world."

He continued, this time taking a fair jab at me: "The Internet and the devices it's spawned are systematically changing our intellectual activities with breathtaking speed, and more profoundly than over the past seven centuries combined. It shouldn't be an act of heresy to ask about the trade-offs that come with this revolution. In fact, I'd think asking such questions would be an important part of the job of a media critic, or a lead Bits blogger."

He even paid me a backhanded compliment, saying that if a Luddite is someone who fears technology, a "Biltonite is someone who celebrates all technological change."

I'm not quite that dogmatic. But his view reminds me of an assessment made in the mid-1990s by Marc Prensky, a software creator who has argued that all kinds of technology should be incorporated into schools and education. To Prensky, there are two camps of Internet users: digital immigrants and digital natives. The natives were born into a world where virtual interaction was commonplace; the immigrants, born before the Internet spread, have had to adapt to its ways.

Over the last five years I've noticed two things that distinguish digital natives from digital immigrants. First, digital natives unabashedly create and share content—any type of content. They aren't satisfied merely having information and aren't at all slowed by doing the creating themselves.

If you have children, you've probably seen the digital natives' creative thinking and need to document. If you watched the inauguration of President Obama in 2009, you will have seen this too. As the president awaited his swearing in, his ten-year-old daughter, Malia Obama, sat behind him taking pictures with her digital camera. There were literally hundreds of thousands of people taking pictures of that event—pictures of Barack Obama would appear on the front page of almost every newspaper and news website around the world—yet his daughter wanted to document the event through her own eyes.

Moreover, digital natives do not distinguish between mainstream stories in the mainstream media such as newspapers and television and those created by their peers. Natives also differ from immigrants in the way they deal with the unbelievable amount of content available to them online.

The digital immigrants came of age reading traditionally packaged information. They felt assured that all the news that was fit to print would be just that: neatly organized, hierarchical, and presented in a specific place on the page. A neat bundle of paper would be on their doorstep when they woke up in the morning, and it would take thirty minutes and a cup of coffee to get through it. Other products existed as thirty-minute TV shows, two-hour movies, and 250-page books. Consumers and creators didn't just switch places. Many people became comfortable with those packages. Packer had a point when he said he felt the need to yell stop. The traditional packages many digital immigrants have grown comfortable with are slowly crumbling.

Now, with so much online, old rules are being smashed and splintered, but new rules are yet to sift out. As mainstream content started to appear online, mainstream packaging didn't quite make the jump. Each day, from 150 to 600 new stories, commentaries, and blog posts are added to a single news site such as nytimes.com, compared with about 150 stories in a newspaper. Mix this with all the sites we see daily and there's too much to consume, with no concise way to organize the ever-growing pile. There aren't any neat little packages.

For digital immigrants—technically, I'm a borderline immigrant—the feeling of being overloaded can be overwhelming. There is simply too much stuff and not enough time to consume it. In recent years, that feeling was compounded as I increasingly was recruited to more social sites by friends and coworkers. Every new interesting blog or website was added to my list of new reading material, and as those digital piles grew, a gradual panic set in—very much like George Packer's anxieties about information overload. I felt like I was being buried alive in an assemblage of words, data, pictures, and status updates. Like E. B. White's digest readers with Irtnog, I wanted—needed—to read everything and felt an inescapable anxiety about missing something important.

The Anchors of Your Online Life

Digital natives, I realized, don't feel this anxiety because they have solved the information overload problem; I have also.

It took me a while, but eventually I realized why: Our social networks are what I call anchoring communities that serve the same purpose for the online world that Benedict Anderson's imagined communities did for the nation. Instead of creating a boundary for a nation, as Anderson's work theorized, these anchors create a boundary in the abyss of the Internet. They help us manage the information overload that traditionalists have come to fear on the Web. Whereas Packer sees an information hell in getting brief updates from dozens of people, I see it in the opposite light: Without my social network to anchor me online, I would be in an information hell.

Nationalism was the glue that held Anderson's imagined communities together and enabled people to think of themselves as Italians, Germans, and Americans. In our digital life, anchoring communities play a role similar to that of nationalism in Anderson's imagined communities.

Why? Because creating anchors helps people feel part of a

community while helping them navigate the digital never-never land. Anchors may seem like just another term for a social network, but they are more than that. The first social networks were not designed to help solve the problems of information overload or to narrow down content; they were meant to be essentially glorified lists of acquaintances—old friends, new friends, friends of friends, and people who used to be friends. Social networks were designed to share status updates, pictures, and eventually news articles. Unintentionally, they have become our online safe havens, our anchoring communities.

Sure, some people still use these social services to tell friends what they've eaten for breakfast, but generally, we've taken the sharing to a whole new level, exchanging expertise and insight and helping one another decide what's important and what is merely digital fluff.

By offering their own digital links and connections, anchoring communities help us cope with the massive numbers of people and the incalculable amount of information online and give us neatly refined selections to sift through together. They help us contain information overflow. These social networks provide cognitive road maps that help us navigate all that information and help relieve the mental taxation of trying to manage excessive information on one's own.

These anchoring and sharing patterns began with America Online's Instant Messenger in the late 1990s. People would copy and paste interesting links of dancing babies, funny animated images, or interesting websites into the instant messages they exchanged with friends and family. Soon those shared nuggets moved to e-mail, then to social networks, which are the boats and moorings of our anchoring communities.

Those broad networks became our content villages. Each individual in these communities brings information to share. Each person decides who visits, who moves in, or who is excluded from their circles. Collectively, we delve into the mass of information.

The social networks help us with this by cutting to the chase. Twitter asks "what's happening?" and Facebook entices you to share "what's on your mind." Granted, sometimes the response is a rather pointless "I need a shower." But when the answer is breaking news about a major event or an intriguing find, our communities may help us start a meaningful discussion and, in many instances, get the information in front of us almost instantly.

Here's how it has changed my experience: For a long time, when I first went to my computer every morning, I would open up a dozen or more different windows to get a handle on what was happening in the world. I had a Google page, nytimes.com, wsj.com, Yahoo!, and so on. The amount of information scrolling across my screen was completely over the top and, often, redundant.

Now, in the mornings, I go to Twitter. There I can look at highlights from whoever I choose to follow. Here's what might come in during the time it took to write this paragraph: My colleague Jim tweeted an update to an earlier story about an oil spill. A friend I met at a conference once, Chris, sent a link about a new blog post on Facebook's confusing privacy policy. My wife sent a link to a foodie blog she's reading. Another co-worker shared a video of Jon Stewart. The links might be from the *New York Times*, CNN, FOX News, individual journalists, friends, random bloggers I've never heard of, or my neighbor.

All of them will sort out significant, interesting, or relevant stories for me, essentially providing my own individual package. I share my finds in the digital flea market in the same way. I still go to specific websites in the morning—the *Times*, Gizmodo, Brooklyn's Brownstoner, and others—and when I find an interesting article from the hundreds I see, I send it to my community in a reciprocal gesture. I don't get paid for it and neither do they, but we help one another navigate the mind-boggling amount of information available on the Web.

Take, for example, an event that took place in my neighborhood. A robber was shot by the police while I was on vacation. My neighbors and friends living in Brooklyn sent up-to-date messages online describing the event almost as it happened. None of these people are reporters or trained journalists, but they were all telling a story and sharing information as if they were on a deadline and receiving a paycheck for their information dissemination.

The Web-Wide World

You might think that these social networks lock us all into a tiny close-minded bubble where we all live in niche silos, unable to see anything but the views that align with the people we interact with online. People who only follow liberals on Twitter will only see liberal views, one might think.

After all, before the Web existed, most of us read a single newspaper in the morning, probably one that was aligned with our political views. We didn't really have the option of reading different newspapers from other locations, either—imagine trying to get a copy of the *Seattle Times* if you lived in New York twenty years ago! It might have taken a week, nothing remotely like the single click that we use today. In the past, the limitation on our ability to see a breadth of options was cost and the difficulty of distribution.

The idea that we exist in a segregated bubble in any community is called homophily—or, in more plain terms, "birds of a feather flock together."

Past research has shown that we tend to align ourselves with like-minded individuals: We sort by income level, age, neighborhood, or similar political or other interests. But on the Web, we see drastically more opinions and viewpoints than we do in traditional media such as television and newsprint.

A paper by Matthew Gentzkow and Jesse M. Shapiro published in April 2010 through the University of Chicago's Booth School of Business argued that the Internet is not only breaking down barriers to different viewpoints but also driving us to see things that we never would have seen otherwise. This is a stark contrast to previous thinking. In 2001, Cass Sunstein, an American legal scholar, penned an article in the *Boston Review*, arguing that our communications were moving rapidly toward a world where "people restrict themselves to their own points of view—liberals watching and reading mostly or only liberals; moderates, moderates; conservatives, conservatives; Neo-Nazis, Neo-Nazis."

But online, Gentzkow and Shapiro found in studying Internet traffic, most news consumers get their information from multiple news outlets—even ones you don't expect they would see: "Visitors of extreme conservative sites such as rushlimbaugh.com and glennbeck.com are more likely than

a typical online news reader to have visited nytimes.com.
Visitors of extreme liberal sites such as thinkprogress.org and
moveon.org are more likely than a typical online news reader
to have visited foxnews.com." After reviewing archival news
data back to 2004, Gentzkow and Shapiro found "no evidence
that the Internet is becoming more segregated over time."

I can tell you firsthand that thanks to my anchoring communities, I see a drastically wider range of viewpoints online
than I've ever experienced reading a print newspaper, watching the nightly news, or reading select niche magazines.

Over the last couple of years, these anchoring communities
have changed the way I receive and share almost every piece
of content and information I consume. My reliance on—and
participation in—social networks and the anchoring community they provide hastened my transition from cable TV to a
computer hooked up to my TV, from a landline for a telephone
to an all-mobile household, and from print books and newspapers to digital readers. I moved to the new systems because
I want everything I encounter and take in to be shareable,
amendable, and receivable.

It's not about watching *Saturday Night Live* on cable TV
versus watching it online; it's that the people I share information with will cut the best clips out of the latest episode and
share them with me. In the same respect, I don't want to be
like my grandmother and clip articles from the paper and mail
them; rather, I want to share the two or three interesting articles I find on nytimes.com each day electronically with everyone who shares news with me.

As a result of this kind of thinking, I no longer feel a shred
of information overload, content anxiety, or fear that I might be

missing something, online or off. Just as those of the print generations feel calmed when their morning newspaper is in hand, I feel confident about my anchoring communities.

The mountain of information available online will continue to grow, and the more information there is, the more likely we are to feel uncomfortable about not having full access to it. No one can possibly hope to eat all the bytes, snacks, and meals being created online. If the thoughtful packaging of information blogs that editors and publishers provide is still too much, our anchored communities will help us manage and edit the overload and provide us with a rich vein of stories.

As these anchors evolve, we will refine them, making important choices about who we believe in and when. At the same time, marketers, search engine providers, politicians, and others will be trying to figure out how to tap into our unique communities to get our attention. How we will know and decide what to trust in the future going forward will become more important—and more complex.

4

suggestions and swarms
trusting computers and humans

The information you get today is coming "more and more through your friends and through your social network. It's being distributed through channels of trust and the trust isn't necessarily the BBC or The New York Times. It's people."
—BJ Fogg

Trust Markets

When I want to know the answer to a simple question—when a movie star was born, what the history of a social movement is, how to fix a technical problem—I put the question in a search engine and "Google" it. More often than not, some of the top listings are from a site such as Wikipedia or Yahoo! Answers or a message board, created not by experts but by people like you and me who want to share their insights and knowledge.

One of the great challenges of this massive, morphing,

growing universe of information is knowing what you can trust and what you can't, even among your anchored communities. This job will get more difficult as clever marketers, technology companies, and others use elaborate computer models to find ways to answer or even anticipate our questions and needs in the future.

Because we come to just about everything with some kind of bias, we may initially trust something solely on the basis of its look or point of view. Liberals may love the *New York Times*'s editorial page but may be appalled at the *Wall Street Journal*'s opinion page, and staunch conservatives may shudder at just the thought of the *Times*'s opinions. Our level of belief and assurance in something determines how we interact, share, and consume it.

In my work at the *Times* and my teaching at NYU, I've been part of many discussions about the value of community-based content found in the likes of Wikipedia and message boards, where the broad Internet community supplies the facts. Even though the wider community is constantly reviewing, double-checking, and revising entries on these sites, many worry—for good reason—about how much you should trust such unknown, unprofessional sources and how you can trust the search engines that take you there.

More and more we are asked to trust a computer, too. Some of the sources we add to our anchoring communities are generated by software bots that use in-depth algorithms to find and highlight interesting news items. An example of this kind of algorithmic reporting is a technology website called Techmeme, which automatically monitors hundreds of technology-related news sources. The site is essentially an ever-changing front page of technology news based on how recently an article was posted, how many times other blogs and news sites have linked to it, and the relevance of the topic on that specific day. Some humans are involved, but only a smidge, monitoring and bringing a little bit of judgment to what appears on the page, and the rest is decided by a computer algorithm. Alltop.com, which pulls together the top stories posted on many different sites, is another aggregator of information. For my colleagues and me, these computer sources are completely credible. I trust the algorithms, often more than statements or claims and press releases sent out by public relations firms, since the computers are seeking information from a variety of reputable news sources. My anchoring communities offer another level of vetting.

Eric Schmidt, the CEO of Google, which handles 65 percent of all Web searches, sees our peers as important in parlaying trustworthy information since we trust them. Schmidt understands that our online communities and their personalized suggestions hold more sway in a search result than does an algorithmic computer search, which is exactly the same for each individual. Just as Foursquare wants to harness your friends' recommendations of restaurants or bars, Google, YouTube, and others hope to do the same for any search result on the Web.

How would it work? Let's say you live in Brooklyn, New York, and you want to find a good Italian restaurant close to the Brooklyn Bridge. You could go to the search engine and type in a search query such as "good Italian restaurant" or "Italian restaurant Brooklyn." You should get the names of many Italian restaurants—but that doesn't mean you're going to find a good meal. The results you see will be the result of looking first for a restaurant in Brooklyn called "good Italian restaurant" and then a mess of other results—yet you won't really know what's good and what's not.

Now imagine that you went to Google and typed in that search query. Instead of an algorithmic answer, your Google results page displayed commentary from people you trust who had eaten Italian food in the area: your friends, family, neighbors, and coworkers, plus whoever else you have designated as trustworthy people in your friend and anchoring communities.

We won't see these kinds of results from Google overnight; the algorithms and artificial intelligence to predict accurately the kind of Italian food you might like still are being developed—but it is getting closer. Making accurate personalized recommendations that are based on your likes and dislikes and the opinions of other people you trust isn't exactly common sense for a computer program to decipher.

The difficultly in making these predictions was highlighted by Clive Thompson, a science, technology, and culture writer who summed up the challenges of making recommendations as "The Napoleon Dynamite Problem." Movies like *Napoleon Dynamite*, Thompson points out, are anomalies in the recommendation functions for Netflix rentals. People either love that movie or they hate it, and there is no rhyme or reason to which of us falls in which category. It's a true anomaly. As Thompson writes, "The movie has been rated more than two million times in the Netflix database, and the ratings are disproportionately one or five stars." People either love it or hate it, and there's no logical answer to their rationale.

Because the movie is so quirky and eccentric, Netflix can't

properly predict how people will rate *Napoleon Dynamite* and therefore can't accurately recommend it to you.

But these services can't afford to get this wrong. If they predict inaccurately, even just once, you may not trust them a second time. If Netflix recommends a movie to you and you really dislike it, the next time you make a rental choice, you're not going to be as inclined to trust the little box that says, "You'll probably like this movie tonight."

Eric Schmidt sees this change too. He has said that Google plans to change its system for filtering and prioritizing your search results over the next five years to accommodate some of the fundamental changes online that are occurring with sites such as Facebook and Flickr that bring together millions of individual viewpoints and insights. "You will tend to listen to other people more," he says. Young people in high school, college, and just beyond are sharing everything and are beginning to enter the workplace. He says they will bring their filtering and community mentality with them into every aspect of their lives over the next few years.

I'm already seeing this happen firsthand. Last year when a friend moved to New York City, instead of buying a guidebook or even searching the Web to find the best area of the city to live in, he created a simple online survey asking about the most important issues for him in finding a new neighborhood and apartment. He sent the survey to thirty or so friends who either lived in New York, or used to, and then used the information to pick his next home. Someday, he might be able to query Google for this insight based on the information his anchoring communities had contributed about their favorite neighborhoods over the years.

Google's Schmidt thinks such a reality isn't too far away, stating that in the next few years, no two Google search results will look the same. If you and I both live in Brooklyn and search for an Italian restaurant, we may get completely different search results, depending on the people in our online communities.

This brings up some interesting questions about how we perceive trust in a digital world. How do we make these decisions with regard to what and who to believe online? If I have a mutual friend online, a friend of a friend, someone I've never even met in real life, do I automatically trust her, too? What happens when I land on a website I've never seen before? How do I know what I'm reading is true and accurate?

So Who Do You Trust?

Traditional media sources build on brands, reputations, and previous experiences to help sell the idea of trust. For example, most people perceive the *Wall Street Journal* as trustworthy when it comes to in-depth stories on the world of finance, even though ownership and management of the paper have changed in the last two years. *People* magazine has the trust of those who want insider gossip about the world of celebrities, and *Wired* magazine has the trust of the technology community on technology trends. But if you took these brands and switched coverage on stories, you probably would see more skepticism. You'd be less likely to trust a *People* article on the latest technology advance in microchips or a *Wired* story speculating on the relationship of Brad Pitt and Angelina Jolie.

Online, however, this sort of media mash-up is already happening. Mainstream outlets, corporations, stores, friends, family, even government, are filtering all kinds of stories and information to you through any number of delivery channels— through social networks, their own websites, and mobile applications. In some cases, they're simply forwarding information from one another. In other cases, the original information may have been forwarded several times. As it all flows in on the same device, with one piece looking just like another, we are challenged somehow to make good decisions about what to believe and what to discard.

So where do we start? Not surprisingly, we tend to trust our friends, family members, and peers deeply. A 2009 Nielsen Online survey of 25,000 consumers in more than fifty countries found that those who participated trusted their friends, family, and peers for advertising and product recommendations 90 percent of the time.

As a rule, we tend to be more distrusting of organizations, news outlets, and government. Over the years, the Pew Research Center for the People & the Press has regularly surveyed the public's views on trust in society. Looking at the trends since the mid-1980s reminds you of the kids' slide at the neighborhood park. The numbers just keep going down. One recent survey showed that between 1985 and 2009, the general public's trust in the accuracy of the news media fell from 55 percent to 29 percent. (Those aren't very reassuring numbers if you make your living reporting and writing news stories.) A separate 2007 study reported that 29 percent of those surveyed trusted large corporations most of the time, although 69 percent trusted them some of the time.

So between our friends and family, our wobbly trust of television and newspapers, and our uncertainty about big companies, there's a lot of available space for others to fill in. Interestingly, people tend to feel somewhat better and more trusting about people they don't know than about the ones they can clearly identify and check out. Another Pew survey asked people in different countries about their feelings of trust toward strangers. In America, 58 percent of those surveyed believed that "most people in society are trustworthy." And although these numbers ranged between 41 percent and 79 percent in other Western countries, on average people tend to trust strangers a little less than 60 percent of the time.

Rick Wilson, a professor of political science at Rice University in Houston, says that numerous research studies and papers show that more than half of society generally trusts complete strangers in an initial interaction. Although he sees people apply dramatically higher levels of trust to friends, family, and peers, he says that our conflicted responses toward politicians and companies have given our online communities more of an opening to win our trust and supply more of our information and insight. That might be the reason we've been so quick to embrace online social networks, he says.

I trust in these complete and often anonymous strangers when I read Amazon.com reviews before I buy a book or when I look at online restaurant reviews before choosing whether to try a new place. True, I don't know who these reviewers are or whether they know what kinds of food or books I like. The restaurant might have deceitfully written some of these reviews— or a competitor might have penned them. But overall, I have developed enough trust in these online reviewers to use their postings in making some general decisions.

Am I nuts to do this? Wilson reassures me that I'm not because I'm not rigid in my assessments. Trust levels continually change, he says, making trust actually something of a game. If I believe what certain reviewers say (and my experience at the restaurant confirms that the salmon is an excellent dish), my trust level rises. If the "outstanding service" turns out to be wretched, my trust falls.

In addition, he reminds me, once someone breaks our trust, it can take a very long time to gain it back—if ever.

Take the website Yelp.com, which allows anyone to write a review of a restaurant or business. The site opened for business in 2004 and grew steadily, winning fans who could find great barbecue on a road trip or the best place to fix a vacuum cleaner. But there were questions from the beginning: How could anyone trust a random person to review a business? What if business owners asked their friends to write reviews or, worse, competitors dumped on one another anonymously? Still, the site generated millions of reviews and became well known for its vast database of locations and opinions on businesses.

Then, in 2009, cracks began to appear in the veneer. Several blogs, business magazines, and newspapers, including the *Wall Street Journal* and the *New York Times*, reported accusations that the company was running something of an "extortion scheme" in which Yelp employees would call the owners or managers of a business and say that they would remove negative reviews for a $300 advertising fee. If the business declined to pay, Yelp might highlight the negative reviews.

In February 2010, a group of businesses filed a class-action lawsuit against Yelp over its aggressive sales tactics. Although Yelp denied the claims, the site's credibility was tainted and many users lost their trust in it. After I wrote about it, one commenter noted, "I believe this about Yelp. I've posted a few reviews on Yelp and for some reason the negative ones never appear (only the positive ones). Since that experience, I've never trusted Yelp reviews again!"

As we start to add and remove people and computers from our anchoring communities, another way to look at our trust in news and other information is as something like a stock market. Each individual or entity within my wide range of networks and connections doesn't receive the same level of trust. Instead, I single them out and apply different levels of authenticity and trust to each person, almost like single stocks in a market. In fact, you can think of it as a "trust market."

Imagine a portfolio of stocks that constantly fluctuate in value. Some ebb and flow, others remain stagnant for long periods, still others rise slowly, and some fall off a cliff. We are constantly applying this mentality to how we trust individuals and the content they deliver within our online communities.

I trust my friends who are news-obsessed to share interesting current events and political stories. I trust my neighbors to share relevant local information, even restaurant reviews. I trust my technology-obsessed friends and colleagues to pass along tech news they find or create. But I wouldn't trust any of them to diagnose an illness or to water my plants, for that matter. They command different levels of trust in my trust market, and they all help me sort through the vast and overwhelming mass of online content. But I also understand that these individual markets can grow and change shape at any given moment.

The shifting nature of trust is one reason I think we're moving toward investing more of our attention and confidence in individuals online and away from traditional companies and their brands. Online, building individual name recognition and trust may be more important than simply affiliating with a trusted institution. For instance, I admire the content in the *New York Times*, but when I go online, I look specifically for media coverage from the columnist David Carr or for simple recipes from the *Times* recipe writer Mark Bittman. I seek out his blog posts rather than his individual newspaper articles, and there I can see his television appearances as well as his columns and read additional tips and suggestions from his readers. After following them for a while, I know that I trust and value their advice.

And it's not just "big-name, big-brand" storytellers who we choose to trust. People like Carr and Bittman have a clear platform for their views, but we're also seeing "no names" build big brands around their big personalities—people who anoint themselves and then build their trust level by delivering content that is valued. If you're an Apple computer enthusiast, you surely will have heard of John Gruber, a Mac expert and writer. He isn't associated with any big-name news outlets or magazines, but he has built a loyal subscriber base with his website daringfireball.com. He is the sole employee and makes a very healthy six-figure income by selling ads on his site and giving talks to companies. Gary Vaynerchuk, a bigger-than-blogger personality, developed Wine Library TV, his own online network of wine reviews and ratings, which claims 80,000 viewers a day. If Gruber and Vaynerchuk can be their own personalities today without the backing of a big-name brand like *Wired* magazine, it's entirely possible that Nick Kristof and Maureen Dowd could still be their own trusted personalities without the *New York Times*. Down the road, I think we are likely to see more reporters and reviewers be known and trusted largely because they have built their own brands, not because of the organization they may (or may not) work for.

Hello, Computer, Would You Like to Be My Friend?

You may not trust a computer algorithm today to tell you where to eat on Saturday night or find a new doctor for you. But eventually you will—and advertisers will try to take advantage of that.

Not all new "friends" in our online communities will be human. More and more, computers belonging to social-networking services, search engines, and maybe media sites will help us sort through the clutter by tailoring information just for us.

Right now, most of the promotions that come to your e-mail inbox or Twitter feed are generic, intended for broad swaths of customers. But as Facebook users already know, advertising frequently is targeted at you, based on your age, your gender, and other information on your profile. A Gmail conversation about dogs may well generate a list of dog-related ads adjacent to your inbox. Search for an address and you'll see local advertising appear right inside the Google Maps. These kinds of smart ads are just the beginning. Even more detailed recommendations are coming that will be based on mathematical formulas and psychological data that will be based on the clicks you make online.

The sites that will provide all this just-for-you data are assuming that you'll be comfortable with a computer knowing a lot about you, just as we've grown comfortable with ATM machines and banking online. In the early days of computerized banking, many people were extremely nervous about trusting a machine with their deposits and withdrawals. A friend recently remembered that her grandmother sat her down when she was a child and explained that "boys and ATM machines just couldn't be trusted." Yet today we use ATMs at delis, on street corners, even inside bank lobbies. Today, there are nearly 400,000 of the machines able to dispense cash, and often they can do more, such as sell stamps or money orders. More often than not, convenience trumps trepidation.

That said, we don't trust these machines or computers any more quickly or blindly than we do real strangers we meet— and we still have a ways to go before we're at a point where these machines are smart enough to offer a normal conversation and allow us to trust them. Just because I'm willing to buy music or a book from iTunes or Amazon, that doesn't mean I'm willing to buy music from any old no-name purveyor with a PayPal account.

Then there's what computer programmers call the cold-start problem. That's what happens when a user doesn't have any information or data in a system, and so the system is unable to make recommendations and we're unable to trust that it really knows something about us. If the computer guesses something about us and gets it wrong, we'll be unlikely to go back to it.

One way programmers hope to tackle the cold-start problem

is to filter and monitor everything about our online actions and our anchoring communities—as Google hopes to do. But these computer systems and online networks are often siloed and separated, too. To solve the problem of digital trust, computers ask people to fill out questionnaires. Some folks simply won't take the time; for others, these surveys make no sense since they ask strange questions, trying to grasp a glimmer of insight into your personality so they can offer better recommendations.

An early study by Timothy Bickmore and Justine Cassell, now at Northwestern University, tried to promote trust in the real estate world by having a computer engage in "small talk." They used a virtual Realtor named Rae, who started conversations with comical banter like "Sorry about my voice; this is some engineer's idea of a natural-sounding voice." After a series of chitchatty questions, Rae began asking more pertinent questions like "What kind of down payment can you make?" and "How many bedrooms are you looking for?"

You'd think that Rae's conversational repartee would make any user feel comfortable trusting a machine, but Cassell and Bickmore found that the results were a little different. Small talk had a much more engaging effect on people who described themselves as extroverts; they felt the machine was more credible and even enjoyed the experience. In contrast, the self-described introverts wanted to get straight to the actual real estate questions and found the small talk annoying. It also limited their trust of Rae. A human being would have been able to distinguish introverts and extroverts, but today, conversations with computers are one-size-fits-all.

BJ Fogg, author and founding professor of the Persuasive Technology Lab at Stanford University, specializes in human-computer interaction and the way we trust machines. Fogg has been exploring trust and machines since the early days of the Web. He believes the issue is not just about trust but about credibility as well. Fogg and his research partner Hsiang Tseng found that in the early days of computing, "the public perceived computers as virtually infallible." Then the assumption that computers were credible quickly started to erode. Fogg points out that credibility in any setting is made up of a variety of different components, including the quality of the interaction, trust, expertise, a lack of bias, knowledge, and experience. Credibility is essentially multidimensional. And since people interact with computers via a screen, that makes building credibility all the more challenging.

When people started to build Web pages, Fogg and his team wanted to understand what made people assign credibility to those pages and trust their content. Since websites were a totally new concept when these studies took place and were a new way to present information, there wasn't much of a starting point. Fogg performed a "large scale credibility study" by showing people different websites that were designed well or poorly and found that what "mattered most was, did the page look good? If it looked good, the assumption was the information was credible, and that was far and away the most important thing that determined whether people thought information was credible."

When I asked Jakob Nielsen, a world-renowned design and usability expert, why people feel more comfortable with well-designed sites, he explained that a lot of the thought process is about comfort and familiarity. "Think about the old banks," he said. "When you walked into the institutions, they had these huge marble statues in the middle of the floor. This was meant to evoke power and strength and confidence so you could trust the institution to look after your money." When it comes to the Web, good design offers the same feeling of trust. Nielsen explained that little things like a logo, a phone number, or clean, well-designed fonts offer a sense of familiarity with real-world objects.

Fogg's research candidly shows that it doesn't matter who makes the information we consume but that we add influence and authenticity on the basis of aesthetics. Or as your mother always warned, we do judge a book by its cover.

I asked Fogg how trust is changing with the next generation of computing and what have become our social networks. He explained that not only will the concept of trust be different tomorrow, the use of the word becomes difficult to apply in new settings.

As an example, he said, "On one hand, trust can mean dependability, like, I'm going to jump off this bridge and here's this bungee cord, and I trust this bungee cord. It's going to be reliable and dependable, and it's going to do what I think it's going to do. Whereas other uses of trust are different. Trusting information or trusting the source of the information goes more towards credibility . . . they're not the same thing," although they have elements of overlap.

That is, we trust our computers to work properly and not explode when the power button is pressed. But whether we trust them to protect our privacy, keep our money or our personal data secure, or even direct us to the right information when we need it is a whole different story. Instead of looking to computers to find the right information or insights for us, he noted that the information you get today is coming "more and more through your friends and through your social network. It's being distributed through channels of trust and the trust isn't necessarily the BBC or *The New York Times*. It's people."

To Fogg, then, a Web page still needs to look nice and be easily navigated to be useful. Now, he says, it matters "who's saying what and if it's somebody I don't know, how many followers do they have?" If it's someone I know, he said, the credibility of that Web page increases dramatically, regardless of design, brand name, or even content.

This isn't to say that elegance in design isn't important. But now there's a human element involved. Indeed, what others think and do has always been influential, and that's something that really isn't changing in the new world. It's just taking on a different form.

So what about those computers? Don't we have to worry about trusting them, too? For now, those distinctions remain relatively separate, and we have the opportunity to make the decision: Do we interact with and trust the algorithm, or do we opt for the human? That option will change. Take the website Wikipedia, the anyone-can-edit encyclopedia. The site employs hundreds of "software bots" that monitor actions on the site, including the creation of new pages or drastic changes to existing articles. If the bots see something out of the ordinary—something they are programmed to look for—they automatically go in and fix the problem. There are hundreds of thousands of these bot-related changes on Wikipedia, and there's no clear distinction between the human edits and the algorithmic ones.

As software and computers become smarter and we start to trust them, we will slowly add them to our trust markets and anchoring communities, both for their dependability and for their credibility. We will have more choices, as we do in deciding between the ATM machine and the bank teller. And more often than not, we're going to opt for convenience, which after all, trumps trepidation.

There is, however, one caveat to all of this: privacy.

It's apparent that the Web and the communities we have joined enable us to share anything from a breaking news alert to the mundane tragedies of our everyday lives. And although we're more comfortable than ever dribbling out short, or long, dispatches from the day, our privacy, or the ability to control it, is still as important as ever.

We can take a look at the social network Facebook to understand just how important this is. It's no secret (both in the media and among its millions of users) that Facebook changes the privacy policy and privacy settings on its website on a regular basis, specifically when it introduces new features to the site. So in early 2010, when the company changed its policy and settings yet again, this time automatically filtering out and linking hundreds of millions of users' information onto the Internet without their complete understanding, there was understandably a tense backlash. Although Facebook was trying to create a better experience for its users, linking individuals' information with their friends and family and in turn creating a social and personalized experience across the Web, the way it went about this completely backfired. I for one didn't want anything to do with the new feature because I didn't trust what was happening to my information, even if it did offer a more compelling Web-surfing experience.

Our online sharing and the mentality of what is private change on the basis of who we let in and who we trust. As a generation comes of age, growing up surrounded by social bubbles online, its members are comfortable sharing publicly with their friends but not with the public they don't know— the general public. If Facebook had decided to launch this new personalized feature with transparency and control, where I understood that my online network was only able to see my actions, I would have embraced it wholeheartedly, but I couldn't consciously share and engage in the public eye without knowledge of which public I was sharing with.

How the Changing Communities Are Changing Us

Now that we've identified how our new anchored communities work and how we build trust in them, we'll turn to how they lead you and you lead them into new directions.

In scientific terms, groups of people can help one another out through "swarm logic." That is, a loose and unorganized group can work together to attack and solve a problem, whether it's hunting for food, avoiding predators, or finding and sharing information.

Another component of this concept is "swarm intelligence." The term was coined by Gerardo Beni, a computer scientist, who theorized that a group can consciously, but more often unknowingly, band together to solve vast and unmanageable problems. Swarms have been used to explain computing, robots, animals, biology, and, increasingly, online social networks. But until recently we haven't really understood how they work, especially with regard to leadership.

In the days of packaged content, the information leaders were the storytellers, such as book authors and newspaper publishers, and those lucky enough to have access to the printing plant. Now the distribution channels matter less, and anyone with an appropriate device can be a storyteller.

But who leads the group in an online social setting? If each person creates her or his own community, isn't there just a chaos of content delivery? Or are there true leaders even in our online social networks? Are we unknowingly developing our own swarms to help manage content consumption?

The way we act online is very organic, similar to the behavior patterns of a species. To see what I mean, let's look at what is known about how fish travel in groups.

In 2008, Ashley Ward from the University of Sydney and a team of researchers, including Jens Krause of the University of Leeds, illustrated how a school of fish will navigate a path on the basis of group leadership.

Ward and a team of biologists took a group of small stickleback fish that usually travel in large swarms and created a lab scenario that included a robotic version of the fish. They placed the fish in a long narrow bath of water and set up two different pathways for them to swim to get from one end to the other. The path to the right had what the researchers called a "predator fish" that was meant to scare and hinder the smaller fish from taking that route, whereas the path to the left was open and free, labeled the "safe route."

When the researchers placed a single live fish in the water, it would immediately swim through the safer route, doing all it could to avoid the predator. But when they added a robotic fish, the live fish would always follow the robot's path, even if it meant going down the predatory pathway. This led the researchers to believe that a live fish will simply follow, even in the face of danger, because another fish has taken a specific route.

To test this, they placed two live fish in the water and set a single robotic fish down the predator path. This time, the two live fish banded together and took the safer route to the left. Leadership was resolved by numbers.

Finally, when the researchers sent two or more robotic fish down the predator path, the live fish—no matter how many there were—would always follow the robots. This led Ward and Krause to believe that swarms make decisions on the basis of a theory they called quorum responses.

Krause explained that in smaller settings, any one fish can become a leader of a group. But when you start to add more elements to the swarm, it takes additional leaders to decide on the direction. Specifically, if there are four or more fish, only two leaders can direct the entire group. Adding a third robotic fish, for example, had absolutely no influence on the direction the fish took. Two was enough to determine direction. Even with small numbers, Krause explained, a collective intelligence sets in.

"Social conformity and the desire to follow a leader, regardless of cost, exert an extremely powerful influence on the behavior of social animals, from fish to sheep to humans," Ward wrote in a research paper on this type of swarm logic.

After the paper was published in late 2008, Krause was approached by a German television station and asked if he was interested in a collaboration to help understand if these theories would apply to a crowd of humans seeking information. He agreed.

As a biological scientist, Krause has spent twenty years trying to decipher the collective behavior, swarm intelligence, and social networks of a wide variety of animals and groups. His studies have looked primarily at leadership within these classifications and tried to explain how hundreds or thousands of individuals can stay organized and share information with such ease and elegance.

With the camera crew in tow, the research team set off to Cologne, Germany, recruited two hundred volunteers, and set up a testing facility in a massive convention center. The basic goal was to "see if it is possible to lead people without them knowing that they're being led."

The study began by placing the volunteers in an empty 90,000-square-foot hall. The participants were told not to talk to one another and asked to move in any direction within the hall but to follow two simple rules. First, they were to move at a normal pedestrian speed, not too quickly and not too slowly. Second, they were told always to stay within an arm's length of any single individual within the group. This request allowed the pack to maintain some level of group cohesion.

Film of the test showed two distinct patterns. First, when a large group is left to wander freely (while still following the basic rules), even without any leadership, it organizes into two concentric circles. This happened every time the researchers ran the test. The groups self-organized to move in a cohesive direction, not dispersing randomly throughout the space. Remember, no one was leading the volunteers and telling them to walk in a specific direction. Still, some sort of organization set in.

Then the researchers secretly asked a percentage of people to try to walk in a particular direction toward a target marked with an X on the floor. The selected people were told to do this while following the two basic rules—move normally and stay within an arm's length of another individual. The volunteers who were asked to walk toward the targets were completely un-aware of the actions of everyone else in the group, including the fact that there were other target seekers.

This led to the second finding, which has become known as the rule of 5 percent. When the small, select group of individu-als was asked to move toward a specific target in the room, the group followed only when 5 percent or more were told to do so. If the researchers told only 2.5 percent of the group to aim for the target, that small group eventually would arrive there, but the other 97.5 percent of the participants would not arrive with them. The rest of the volunteers managed to stay within the concentric circles but not follow the folks seeking out the X on the floor. But as soon as the researchers upped the number to 5 percent or more, the entire crowd of two hundred ended up following and everyone made it to the target.

In an interview, Krause explained that the smaller groups' goal was not just to wander but to walk to the target while stay-ing with a group. It became a "self-organized process because nobody has knowledge of what the group is collectively about, or what the individuals all know. Everybody is just following his or her local route. So as a result, we see collective locomo-tion towards the target."

This theory applies whether you have 5 percent, 10 percent, or even 50 percent headed in one direction. The whole group will always reach the target if 5 percent or more knowingly or unknowingly lead the way.

The rule of 5 percent becomes increasingly important in settings in which a group shares information about a predator or food. Online, in the absence of both predators and food, we collectively avoid substandard, inaccurate, or ineffectual content and seek premium, quality information. Krause believes that these tests show that when a "few individuals, or a small proportion [of a group], receive information that the others don't have, then they can become disproportionately influential" within a group. When you apply these findings to our online experiences, they illustrate how anyone, regardless of background or expertise, can become an influential individual within a group.

Krause believes that when we all have the ability to share data, the information sharing becomes completely egalitarian. If you have distinct information at a specific moment, you will become the temporary leader of the group, with the ability to influence the flow and formation of the swarm.

There's another important component to the ebb and flow of sharing and leading. Online, just as in these real-life studies, positive feedback plays a key role. "An individual does something that is copied, and the more individuals copy it, the stronger the urge becomes for others to follow suit," he said. If you imagine a swarm of insects floating through the air back and forth, or a school of fish, or even the people in the testing facility in Germany, they move in an elegant swooping circular pattern as the group's leaders change and gather new information.

Something similar can happen online with the information we share and consume. Any single individual can find something interesting and send it to the group, and if it's stimulating and appealing, they in turn share it with their community. The cycle of mass of content seekers will gravitate toward the X on the Web. Then the pattern begins all over again.

If the news is that important, it will find me.

—A college student explaining his news habits in a
  focus group

So are we really nothing more than a school of dull fish? Could anyone with an audience of 5 percent lead and change an entire online group? Couldn't a smarmy individual go off into one online world or another, recruit a few hundred people within a network, and drive you to click on a link?

Luckily and happily, no, because in real life in the online world, we aren't all locked up in a big hall with our arms nearly touching. Each community we set up is individualized to each of us. For instance, you're the queen bee of your own anchoring communities, the person who filters the information you most want. But by belonging to a social network, you're also a worker in someone else's hive. Since no two social groups are alike, the whole group becomes notoriously difficult, if not impossible, to control.

Most often, you're not actually leading the group. You're simply a part of the information sharing, harboring a collective intelligence. You may decide who enters your own network, accepting friends' requests or following someone's actions online, but you don't control what they share and consume. You just decide if you're going to pay attention to them.

You also don't seek out the same information as others in your communities. Your queries and interests are based on information channels that differ from the ones I use. Yet if Sam H. and others in my Foursquare world start to rave about a new restaurant, I probably will check it out. If several Twitter friends tell me about a great story or share breaking news, I will pay attention. My anchoring communities will bring the news or their discoveries to me, helping me sort, filter, and distribute a living and ever-changing stream of information and experiences.

A yearlong research study released in April 2010 by researchers from the Department of Computer Science at the Korea Advanced Institute of Science and Technology used the social-networking service Twitter to explore the theories of social news gathering and dissemination further.

In July 2009 the researchers set up twenty computers to suck in every single piece of information shared on Twitter: every tweet, every retweet (when someone resends a tweet from another user), the number of followers, and so on. In their collection the researchers gathered 41.7 million user profiles, 1.47 billion social relations, 4,262 trending topics, and 106 million tweets.

So what did they find with this treasure trove of data? A majority of the conversation taking place on Twitter at the time was about news and information sharing. Looking through Twitter's trending topics during this period, the researchers found that more than 85 percent of the top-level topics were headline news or something newsy in nature. They also found that no matter how many people follow a user on Twitter, anything that is retweeted by other users will reach up to 1,000 users on average.

We can get a real-life glimpse of content spreading through these anchoring communities in a piece of research from a project by Gilad Lotan, a developer and researcher at the Microsoft Research labs in Cambridge, Massachusetts.

In June 2009, as the Iranian revolution spread across the Web, *The Nation* magazine wrote, "Forget CNN or any of the major American 'news' networks. If you want to get the latest on the opposition protests in Iran, you should be reading blogs, watching YouTube or following Twitter updates from Tehran, minute-by-minute."

As this online revolution took place, Lotan built a tool to monitor how news spreads on Twitter, aptly calling the project ReTweet Revolution. He monitored the use of Twitter over ten days in June as the rebellion against the rigged election was taking place in Iran. Lotan sifted through 230,000 tweets and found 372 distinct threads of information around the protests in Iran. As the Iranian government tried to suppress the spread of information on the Web, shutting down websites, information was able to slip out of Iran only through a few individuals on Twitter. One was a student who called himself "tehranbureau" on the social network. As the protests started to unfold, Lotan said many of the Twitter users were reaching very few followers, even though many of those sharing the news inside Iran had only twenty or thirty people following them. But as people all over the globe shared the news by retweeting it, the content eventually was seen by tens of thousands. That in turn influenced the coverage by mainstream media, something that until that time would have been highly unusual.

This swarmlike behavior does more than spread important news. It also dissipates our fears of information overload or the converse, that we might be missing something. When members of my anchored community let me know that certain products are worth consuming, I trust their recommendation because the social networks I've set up have been selected by me and cleansed of members I don't trust—whether they are computers or people. I'm sure that in some instances I've been flushed from someone else's trust market as well.

This new way of consuming information and storytelling online doesn't bode well for individuals or companies that create mediocre content and cookie-cutter storytelling. The new mentality says that if it's not good or important, the group won't share it. Furthermore, it no longer matters who created the content; if it doesn't satisfy us, we're not going to share or filter something up the food chain.

During the 2008 presidential election year, Brian Stelter, a Times media reporter, found that people twenty-five years old and younger tended to share political news with their groups of friends through e-mail or other social outlets. They provided news and information to their friends and relied heavily on their friends to do the same. They didn't need to go through all those newspapers and magazines looking for unexpected stories to find the significant stuff. Their friends did that for them. They used these anchoring communities, their own personal public spheres of friends, family, news outlets, blogs, and random strangers—people like Sam H.—to share and disseminate.

This is the way I navigate today as well. If the news is important, it will find me.

Maria Popova runs the blog Brain Pickings, which looks for fun and interesting tidbits online. She calls herself a cultural curator, and she searches for interesting cultural references on blogs, websites, and Twitter feeds and then shares them with thousands of strangers who follow her, passing along the best of the best on her site and through her Twitter account. She calls the process "controlled serendipity": "I scour it all, hence the serendipity," she said. "It's essentially 'metacuration'— curating the backbone, but letting its tentacles move freely. That's the best formula for content discovery, I find."

As with porn, whether the content is produced by a hundred-million-dollar studio or by people in their bedrooms with a webcam, good content will rise to the top, and our special communities and collective intelligence will help get it there. Our trust communities will help us filter the tsunami of data, opinion, insights, news, and reviews coming our way so that we feel neither overwhelmed nor anxious.

Standing on the sidelines of these social networks and trying to figure out what they all represent and if there's a purpose to these experiences can be downright daunting. I completely empathize with the trepidation of George Packer and others and their rightful concern that there are only so many hours in the day to deal with too much already. I've been there, and although there is some transition in the process, I'm convinced that being guided online by communities that I trust won't create an information hell that leaves you gasping for air. Instead, trusted anchoring communities will help you filter and navigate a bigger world in an eye-opening way that has never been possible before. You just have to get your brain around the possibilities.

5

when surgeons play video games

our changing brains

Men were twice as likely [as women] to tweet or post status updates after sex.

This Time, We're Really Going to Hell

In the summer of 2008, Nicholas Carr, an author and writer for *The Atlantic* magazine, felt his brain slipping ever so slightly from its moorings. In the past, he wrote, "Immersing myself in a book or lengthy article used to be easy."

Not anymore. "Now my concentration often starts to drift after two or three pages. I get fidgety, lose the thread, begin looking for something else to do," he said. "I feel as if I'm always dragging my wayward brain back to the text."

The problem, he concluded, was the Internet generally and

Google quite specifically. In a piece titled "Is Google Making Us Stupid?" and later in the book *The Shallows: What the Internet Is Doing to Our Brains*, he frets that having snippets of massive amounts of information right at our fingertips may be eroding our ability to concentrate and contemplate.

Déjà vu all over again, don't you think?

In fairness, Carr recognizes that the printing press caused similar hand-wringing. And even though some of the predictions came true—the press actually did undermine religious authority, for instance—the many advantages of printing far outweighed the concerns. So he admits that he may be wrong and "a golden age of intellectual discovery and universal wisdom" may emerge from the text, tweets, bytes, and snacks of today's online world. But he still worries that deep thinking and serious reflection will be forever lost in the data stream of information the Web affords.

Although Carr is fatalistic about the future, his balanced, well-researched article offers a thoughtful perspective. Most of those who are skeptical about the shift aren't so reflective. In a *San Francisco Chronicle* article headlined "Attention Loss Feared as High-Tech Rewires Brain," the author, Benny Evangelista, citing some mental health experts, saw interpersonal relationships breaking down and attention deficit disorder increasing as more people found themselves unable to separate from e-mail, Facebook, and Twitter.

How bad is the attention loss? The inability to detach oneself from electronic updates has spread from the office to the restaurant to the car—and now has reached into the bedroom. The story quotes a survey that found that 36 percent of people age thirty-five or younger used Facebook or Twitter after having sex. "Men," the story noted, "were twice as likely [as women] to tweet or post status updates after sex."

Said an executive who commissioned the survey: "It's the new cigarette."

Other news stories and books drip with angst over how these new technologies may be destroying us, ruining our intellect, quashing our ability to converse face-to-face, and fundamentally changing relationships for both kids and adults. "Antisocial Networking?" a *New York Times* story asked, questioning whether time online diminishes intimacy and destroys the natural give-and-take of relationships. "Scientists warn of Twitter dangers," said CNN.com, stating that researchers had found "social-networking tools such as Twitter could numb our sense of morality and make us indifferent to human suffering. A number of books, such as *The Dumbest Generation: How the Digital Age Stupefies Young Americans and Jeopardizes Our Future* and the previously mentioned *Distracted: The Erosion of Attention and the Coming Dark Age*, add fuel to the fire.

But it doesn't stop there: There's a recent, often-quoted study, "Emails 'Hurt IQ More than Pot.'" A survey of more than a thousand Britons found that the IQ of those trying to juggle messages and work fell ten points, more than double the drop seen after smoking marijuana.

Over and over at speeches and conferences, I hear the same kinds of fears and anxieties that new technologies and developments have generated for decades: Our brains weren't wired for all this fast-paced stuff. We're too distracted to do meaningful and thorough work. At the same time, our entertainment is also dangerous and damaging, people tell me. Video games will destroy our children's brains and their relationships—if

Twitter and Facebook don't do so first. We cannot effectively multitask or jump from e-mail to writing to video, and we never will be able to.

There may be some truth to some of this; we may well be fundamentally different when this is all over. But for the most part, I believe it's bunk. Just as well-meaning scientists and consumers feared that trains and comic books and television would rot our brains and spoil our minds, I believe many of the skeptics and worrywarts today are missing the bigger picture, the greater value that access to new and faster information is bringing us. For the most part, our brains will adapt in a constructive way to this new online world, just as we formed communities to help us sort information.

Why do I believe this? Because we've learned how to do so many things already, including learning how to read.

We were never born to read.

—Maryanne Wolf, *Proust and the Squid*

Some argue that our brains aren't designed to consume information on screens, or play video games, or consume real-time information. But the same argument holds true for the words you're reading now. It's true: Your brain wasn't built to read. Several thousand years ago, someone created symbols, which ultimately became an alphabet. That alphabet formed into a written language with its own set of unique rules. As a result, the organization of the human brain changed dramatically. But the human brain doesn't come automatically equipped with the ability to read these symbols. It's something that has to be rewired in the circuitry every time it happens. Our brains are designed to communicate and to tell stories with language, whether that is with clicks of the tongue between indigenous tribes in the rain forest or with the English language. But reading letters and words is essentially man-made, just like video games and screens.

Even today, when children learn their letters and form them into words and sentences and big, powerful ideas, their brains still have to re-form and readapt to make the information fall into place.

Stanislas Dehaene, chair of Experimental Cognitive Neurology at the Collège de France, has spent most of his career in neuroscience exploring how our brains learn to read and calculate numbers. He explains that human brains are better wired to communicate by speaking. In the first year of life, babies begin to pick up words and sounds simply from hearing them. Sure, they need some help identifying that a cup is a cup and Mommy is Mommy. But by two years old, most children are talking and applying labels to objects without any special lessons or drills.

This is not the case with reading. Most children, even if they share books with their parents and hear stories every single day, won't pick up reading on their own. Instead, they must learn to recognize letters one by one and put them together into sounds or words before recognizing whole sentences and thoughts. They must learn to decode the symbols.

Some research suggests that in doing this, children and even adults actually develop a new area within the brain. Manuel Carreiras at the Basque Center on Cognition, Brain and Language has taken research on language into other complex areas. Carreiras's work over the years has been focused on the neural processes of human language and the way humans comprehend differently when reading and when interpreting sign language. When he wanted a better understanding of how people learn to read, he decided he needed to find illiterate adults to see how their brains adapt before and after learning how to read words.

At first Carreiras had trouble finding a group of adults who really didn't have any reading skill, but finally, he recruited forty-two veterans of the Colombian guerrilla wars. Twenty of the ex-fighters had recently completed a Spanish literacy training program to teach them to read. The other ex-fighters still needed to take the course and were for the most part illiterate. The former fighters were tested, taught to read, and then tested again. In the process, areas of the brain actually grew and formed connections that had not existed earlier. The brain had rewired itself while the guerrillas learned how to read.

Carreiras found that the brain changes its structure when someone properly learns to read, particularly in the white matter, which creates connections and helps information move between different areas of the brain. He explained, "We found that the literate members of our group had more white matter in the splenium . . . a structure that connects the brain's left and right hemispheres—than did the illiterate members." As the former guerrillas learned to read, the scientists used imaging techniques to measure what was happening. They saw that reading triggered brain functions in the same areas that had grown over the course of the study. In other words, even adults were able to create new neural pathways as they learned a difficult new skill.

What's significant in this example is that our brains are

something like a muscle, which can grow stronger and more powerful with practice and work. Today, technology is building new connections as our brains interpret content and receive stimulation. There's a constant and simplistic iterative adaptation taking place in our brains as we use our computers, mobile phones, and e-readers. Our brains are learning how to navigate these gadgets, just as they do when we learn how to read.

There's one piece of this puzzle that's important to point out. With the use of computers and digital technologies, our brains are not evolving. Human beings evolve at much slower pace than do new communication mechanisms and the technologies we invent and create. Neuroscientists I spoke with explained that a brain from five hundred years ago or even ten thousand years ago will look pretty much the way it does today, just as humans look pretty much the way they did a few thousand years ago.

To illustrate this point, let's hypothetically travel back two thousand years and find a newborn baby. Imagine taking that baby and transporting him through our time machine forward to today. This child would be raised in our technology-rich society, growing up in a world of iPods, video games, the Internet, mobile phones, GPS, robotic Elmo toys, banner ads, and more. I asked several neuroscientists if this baby born two thousand years ago, I was told, would likely grow up differently than would a child born today. The resounding answer was "no." A newborn's brain from two thousand years ago, I was told, would likely look and work exactly the same as a brain does today.

But what if you took an adult—let's say a thirty-year-old man from two thousand years ago—and dropped him in the middle of Times Square. He might well experience a panic attack from all the crowds, cars, flashing lights, and stimulation. But, neuroscientists said, his brain would begin to adapt. He might never get to a point where he could talk and simultaneously send text messages, but numerous research studies show that our brains are capable of substantial adaptation in about two weeks and in some instances seven days. Our two-thousand-year-old man would be just fine. His adaptation to society and the new stimuli would just take brain training, and not as much as you might think.

How do our magnificent minds adapt? In 2008, a group of neuroscience researchers from UCLA's Semel Institute studied the brain activity of twenty-four volunteers when the subjects were reading a book or surfing the Web to see if the Web was rewiring the way our brains function.

The volunteers were divided on the basis of how much experience they had using computers and the Internet. Twelve of the participants were labeled "Net Naive" because they used the Internet or computers once a month at most. Asked to rate their tech savvy, they gave themselves a rating of minimal to none. The other twelve participants were labeled "Net Savvy." Those in this group used a computer at least once a day, and most of them were online numerous times throughout the day. The members of this group considered themselves moderate to expert on computers and the Internet.

The researchers showed volunteers different types of content while monitoring them using functional magnetic resonance imaging (fMRI) scanners, special machines that allow the research subjects to watch screens or perform certain tasks while the scanner records the blood flow in their brains and how the brain handles its processing.

First, volunteers were shown a table of contents from a book and given fifteen seconds to pick the chapter they wanted to read. Then they had just under thirty seconds to read a page of the book. Next, the same participants were shown a search page from Google and asked to decide on a search and enter a word in the search box within fifteen seconds. The display took them to a website that corresponded with their search, and they were asked to read the page for an additional thirty seconds. To make sure they were paying attention, the participants were told they would be tested on their reading of both the print and digital versions.

When reading the printed page, the Net Naive and Net Savvy brains reacted the same way. Those brains were slightly stimulated, although there was a little less activity in the brains of the Net Savvy while reading the printed text. But during the online searching and reading tests, the Net Savvy brains were much more active. In fact, the Net Savvy group showed almost twice as much activity while online compared with reading a book. The reading task stimulated parts of the brain used for language and reading, memory, and visual abilities. In comparison, the Web surfing task activated the same areas of the brain as reading, but in addition, the brain was involved in decision making, complex reasoning, and vision detection.

Even more interesting, the volunteers weren't a bunch of young kids with malleable brains. Rather, the group consisted of people between fifty-five and seventy-six years old, all of them digital immigrants with varying degrees of success in adapting to the online world. The Internet wasn't even around in a meaningful way until they were in their late thirties or early forties to midfifties, yet the brains of the Net Savvy rewired and sprang into action to work with this new stimulation.

What was happening with those brains is a process called neuroplasticity, a theory that our brains' 100 billion neurons, or nerve cells, can re-form, or create new cells and new connections, as we learn and grow.

Many new activities we engage in on a daily basis can make this happen, from touching something hot for the first time to using the Internet or even juggling, as Bogdan Draganski and a group of scientists from the department of neurology at the University of Regensburg, Germany, discovered.

Draganski, with previous brain research as a foundation, developed a hypothesis that our brains must act differently when they learn something new. After watching a group of kids text messaging on their mobile phones at dramatic speeds, he wondered if sending hundreds of messages a day with one's hands made the thumbs work differently. He theorized that the brain correlations that operate these functions should look different from what they look like in people who rarely text.

To explore this theory further, Draganski told me in a phone interview, he got permission to scan the brains of a small group of young people. The initial results showed that the heavy texters had a larger area of mass in the portion of the brain that controls the right hand, but other areas were similar to normal brains he had studied earlier. Draganski believed that this larger mass most likely signified heightened use of the right hand used for texting. His original goal was to understand if brain growth would become more obvious over time as more kids learned to text. But, he said in an interview, with so many young people already familiar with texting, he decided to switch to a different task that involved a clear and steep learning curve: juggling.

Draganski and his researchers took a group of participants who had never juggled before and measured the gray
matter, the neurons, in each of their brains as they gradually
learned to juggle with three and then four balls. As Draganski predicted, he saw significant areas of growth of gray matter in certain areas. The motor areas of the brain actually grew
over a three-month period of learning. When the participants
stopped juggling, however, their gray matter began to recede
and return to its previous size and shape.

Another group of researchers in a different study found that
when a completely new task is learned, changes in brain shape
are visible after a mere seven days of practice.

When these theories were tested in a later study by the
UCLA researchers in late 2009, it was found that Net Naive
Web surfers could catch up to the Net Savvy. When the Net
Naive repeatedly used the Internet over a one-week period, the
brain scans showed that they, too, started to adapt and respond
to the online experience in a very similar fashion to the Net
Savvy. Their brains also began to show twice as much stimula-
tion from reading a Web page as from reading a printed page.

Gary Small, director of UCLA's Semel Institute for Neuroscience & Human Behavior and one of the nation's leading
experts on memory and aging, was one of the key researchers
in this study. He said the brains were learning, benefiting from
practice and experience. In theory, Small said, as we learn, the
brain should show less activity. For example, when we get a
new phone, it takes a while to figure out where all the functions
are hidden. "At first I'll show a lot of activity in my brain," he
said, but then, once he gets used to the experience and better at navigating the device, the activity should slow down. At

that point, he said, the brain's "synapses . . . will grow, become strengthened, and then become efficient," and less activity should be required.

But that isn't what happened when he watched people become experienced digital surfers. Instead, his research found, our brains work completely differently while reading online than while reading a printed page, making numerous decisions based on the many options, menus, photos, text, and links on each page. In fact, the first study concluded, "Internet searching appears much more stimulating than reading."

For More Details, Click Here . . .

What happens when we're online that keeps the brain so busy?

The online experience isn't simple or controlled; it's like the Wild West. The user interface alone is enough to send you running for the comfort of the printed page. Every last piece of real estate on the screen is vying for your attention. Your Web browser has back buttons, reload buttons, and a bright red stop button that screams, "Hey, look at me!" Other windows may be floating in the background of your computer screen. You probably have a desktop image of your cats or a cute baby.

Then there's the actual Web page, which includes eye-popping banner ads, search boxes, logos, and colored text showing you links to other Web pages, which then link to even more Web pages. In the course of a day, you might go to a few news websites, read a blog or two, look at the weather, search Google for a range of answers, and buy a book on Amazon or eBay. Before you know it, you may have visited well over a hundred Web pages in a day. That may not seem like a lot, but the amount of content you see can be mind-boggling.

In research at the *New York Times* labs, we found that on average, each Web page in the top 100 most-visited news and information sites and blogs online has about 370 links; some have more, some a little less. So if you were to visit the home page of each of those top websites in one day, you would have faced more than 37,000 links.

It can be completely overwhelming for our brains to navigate the Web. No wonder Small's study said a book is sometimes less stimulating than the Internet. The Web is vying for our attention constantly.

Although our anchoring and trust communities help us determine where to go and what we can trust, the links also help us navigate the trails. Imagine what it's like when you walk into a large bookstore like a Barnes & Noble. You will see thousands of books displayed on the store's shelves. All around, there are filters to help you find where you want to browse and what you want to buy. Books are organized by subject matter. There are recommendation tables organized to help you find specific genres. There are top ten lists, top fiction lists, employee recommendations, *New York Times* bestsellers. Or you can base your decision about what to read on a friend or coworker's suggestion.

The Web eventually will get to this place, too, and once again, history can show us the way. A front page of the *New York Times* a hundred years ago was a messy cacophony of more than sixty headlines and stories. Today's paper has a grand total of six stories. You would think that over a hundred years, during a time when content creation has essentially exploded into trillions of little bits of information, the newspaper would stuff in more stories and headlines. But the *Times* and other newspapers came to understand that their job wasn't to print every single piece of news of the day but to do a better job of filtering it. It's an editor's job to reduce what a reader's brain has to wrestle with.

Until now, the Web has taken a contrasting path. When we go online, the limitations of the print experience—the size of a piece of paper—just evaporate. In 1995, when the *New York Times* debuted its website, the concept was to re-create the newspaper experience in a digital form. On the home page, you might have seen a whopping two stories, with one photo and links to eighteen different sections of the website. That's a grand total of close to fifteen links on the home page. Fifteen years later, the home page of nytimes.com has more than 550 links, nearly 300 of which are headlines related to stories. No wonder the brain is busy.

Beyond the links, the websites have plenty of words, too. In a story I wrote and visualized for the United Kingdom version of *Wired* magazine, I found that the top two hundred news and information sites in the United States and the UK put forth an astounding grand total of 487,881 words and 66,248 links. And get this: Hitting those two hundred sites is the equivalent of flipping through Leo Tolstoy's *War and Peace*, which is 480,000 words long.

Granted, two hundred Web pages is a lot. If you're flying along like that, you're probably a media junkie on a tight deadline or a bored Web surfer stuck inside on a rainy day. But let's go a little further and combine this with all the information we eat up every day.

Technology researchers at the University of California–San Diego looked at the number of words we read in all kinds of media and quantified it as if it were placed on a hard drive at the end of each day. The researchers calculated that in 2008 American households collectively consumed 3.6 zettabytes of information.

What's a zettabyte, you ask? I had to look it up too. Here's how I described it in a blog post for the *Times*: "I'll be honest: this is the first time I've ever used the word zettabyte. I've heard of petabytes and even exabytes, but zettabytes are a whole new level of bytes. If a zettabyte is beyond your comprehension, too, it's essentially one billion trillion bytes: a 1 with 21 zeros at the end. To put that into perspective, one exabyte—which equals 1/1000 of a zettabyte or 1 billion gigabytes—is roughly equivalent to the capacity of 5.1 million computer hard drives, or all the hard drives in Minnesota."

In other words, it's an ocean of information. The researchers also found that the average American can take in up to 36 million words a year. This doesn't mean that we are reading 100,000 words every day. But it does mean that we are subjected to these words through any number of channels: television, radio, text messages, the Internet, video games, and advertising.

There's no sign of this slowing down. The researchers also calculate that the information wave continues to grow 6 percent each year, representing a 350 percent increase since 1980 in the amount of information we regularly confront.

Finally, our brains are stimulated while we're online by the physically interactive and unpredictable nature of using a computer: You're challenging your brain by holding a mouse, looking at a screen, and navigating through choices and buttons. It's a very hands-on experience that is completely different from the more passive and linear activity of reading a book or watching TV or a movie. When you're reading or watching a movie, your body and your hands are relatively still. And though you can certainly skip around, you're more likely to read or watch from the beginning to the middle to the end.

Although there is a beginning, middle, and end to most online content, those links also form thousands of branches of information that essentially allow you to devise your own narrative, creating a whole new form of storytelling. The Web does have its linear possibilities, but you have to add a broad level of multidimensionality. There are uncountable coexisting narratives and stories.

All of this is enough to make your head spin. It makes perfect sense that our brains are active as a scanner when we're online.

Online our brains are stimulated, calculating and exploring. They are working differently. This is consistent with another development researchers have found: Mastering another electronic challenge—video games—also engages the brain and may actually make us more adept at certain tasks.

But this doesn't mean our brains can't handle this new form of storytelling. It just means we are telling and consuming stories differently. In addition, content creators and consumers are feeling their way through a digital metamorphosis. It took decades for the editors at the *New York Times* to realize it wasn't in their best interest to put sixty headlines on the front page of the newspaper and that in actuality it made more sense to put six highly curated headlines on that page.

As our brains adapt and continue to grow and change shape, the technology and storytelling will continue to do the same thing. Our brains have done this successfully for thousands of years as they have learned new forms of communications and storytelling.

Does Your Surgeon Play Video Games?

The next time you have surgery, ask your surgeon if he or she played video games in the past.

A few years ago, researchers quizzed more than thirty surgeons and surgical residents on their video-game habits, identifying those who played video games frequently, those who played less frequently, and those who hardly played at all. Then they put all the surgeons through a laparoscopic surgery simulator, in which thin instruments akin to extremely long chopsticks are inserted into one or more small incisions through the skin along with a small camera that is inserted into an additional small opening. Minimally invasive surgery like this frequently is used for gallbladder removal, gynecologic procedures, and other procedures that once involved major cutting and stitching and could require hours on an operating table.

The researchers found that surgeons or residents who used to be avid video game players had significantly better laparoscopic skills than did those who'd never played. On average, the serious game players were 33 percent faster and made 37 percent fewer errors than their colleagues who didn't have prior video-game experience.

The more video games the surgeons had played in the past, the better their numbers. This wasn't tested on a group of kids who played twelve hours of video games a day and hadn't showered in weeks. These residents and practicing surgeons simply played three or more hours of action video games a week. Some of the more advanced video-game-playing students managed to make 47 percent fewer errors than others and were able to work as much as 39 percent faster.

The results were surprising given the criticism video games have received for rotting young minds, turning upstanding youngsters into juvenile delinquents, and just wasting time. Instead, surgeons and researchers have begun to test whether the games should be a key part of a future surgeon's education, since speed and accuracy are crucial to conquering the learning curve associated with using laparoscopic techniques to perform delicate procedures. Game skill, the researchers theorized, could translate into surgical skill and help cut "medical errors," which have become the eighth leading cause of death in this country.

A couple of years ago, a researcher at Arizona State University tried this out on surgeons at Banner Good Samaritan Medical Center, using a Wii golf club that was reshaped into a laparoscopic probe. One group of residents played a suite of games called Wii Play and a game that involves subtle hand movements, Marble Mania, using the probe, while another group didn't. The game players showed 48 percent more improvement in performing a simulated laparoscopic procedure compared with those who didn't play.

But not every game helps surgeons improve their skills. It turns out that Wii's Marble Mania stimulates the areas of the brain needed for surgery. Games such as Wii Tennis, where you swat your arms in the air as though you were hitting a virtual ball, did not help surgeons' scores. But many studies have found that even limited practice on video games may increase speed and skill in surgery.

It's no surprise, of course, that dexterity improves with practice. But what makes these studies stand out is how effectively human brains can make the leap to conquering new technologies and then putting those new skills to use in innovative and varied ways. For example, these studies consistently show that playing video games improves hand-eye coordination and increases one's capacity for visual attention and spatial distribution, among other skills. These increased brain functions are tied not only to game play but to several other real-world scenarios, including surgery.

You may feel like your brain cannot cope with so much information or jump seamlessly from one medium to another, just as you may have felt in high school that you couldn't learn a foreign language or conquer higher math.

But as the brain faces new language (or acronyms and abbreviations), new visual and auditory stimulation, or new and different ways of processing information, it can change and grow in the most remarkable fashion. In fact, it may well be a natural part of human behavior to seek out and develop unnatural new experiences and technologies and then incorporate them into our daily lives and storytelling.

Seventeen Buttons and Ten Fingers

I personally can't justify playing video games to practice my surgical skills. Medical procedures aren't exactly my strong suit. But video games have helped my brain master new forms of storytelling in ways I didn't even realize.

The first video-game system I owned was the Atari 2600. The Atari debuted in 1977 and made its way into my home when I was five years old, in 1981. I don't remember a lot about that year, but I do remember holding that Atari joystick in my clammy little hand, excitedly batting a square, grossly pixelated ball across a screen with my friends. I played games like Pong and Space Invaders. Today they are artifacts of game history, but at the time they stimulated my mind to no end.

The Atari's game controller was a simple, almost primitive device. On the top of its square-shaped paddle was a single joystick. The upper-left-hand corner was home to a single orange button. That was it: one stick, one button.

Today, a single controller for the video game system in my living room has fourteen buttons and three multidirectional joysticks. I still have only ten fingers, but on today's controllers there's a place for seventeen different fingers—not including the indisputable fact that I actually have to hold the controller. Yet when I sit down to play video games, I don't fall into a panic or become overwhelmed with all the buttons and joysticks. I just play. My brain and the technology have both adapted.

A multitude of studies, going back thirty years, show that my experience isn't isolated—video games are actually extremely stimulating to the human brain.

One of the earliest and most famous studies took place in 1991, when Richard Haier, a psychologist at University of California–Irvine, studied the newly released game Tetris. Haier recruited a group of participants between the ages of nineteen and thirty-two who had never played Tetris before. Over an eight-week period, the participants were asked to play Tetris twice a week. Then, before MRI scanning was a possibility, they were asked to go through a positron emission (PET) scanner, which measured glucose levels and pathways within the brain to learn where oxygen was being used and see where the brain was being stimulated.

Haier remembers that it was easy to find a group of students for the study who had never played video games before. "It was the early nineties and not many people had heard of Tetris yet, so it was easy to recruit new players for the study," he said.

At the start of the study, Haier and his research team explained what Tetris was and how to play, and then the participants' game scores were monitored as they played. At first their scores were extremely low, just five or ten points each time they played. But numerous areas of their brains showed a dramatic increase in activity in many different functions. Haier explained that these data showed the games were extremely stimulating to the players.

As the study progressed, the participants' game scores increased pretty dramatically, climbing to more than 100 points per game. But as the scores increased, the amount of brain stimulation or activity decreased, to the surprise of the researchers. The PET scans that previously showed glowing activity returned muted levels of stimulation for many parts of the brain, although other parts remained busy. The brain had adapted fairly quickly to this new visual and interactive form of storytelling.

Although Tetris involved numerous different brain tasks,

hand-eye coordination, spatial observation, planning, sight, sound, and more, the brains of novice players quickly figured out how to master each of the jobs.

Haier pointed out that the brain benefits from stimulation. "Clearly video games are impacting our brains—this is why people are so emotional about them," he explained. "However you are engaged in the world, the brain is engaged. This is why parents buy toys for the crib, things that move around and make noise." Haier reiterated that "the brain is very adaptable and every generation basically has new stimuli that previous generations had not encountered." Video games are not bad for our brains, he said; they are just new, and our brains need to figure out how to use them.

The players' rather rapid improvement in Haier's studies reflects the concept of plasticity, the way our brains change as we learn new things. In a theory originally put forth a century ago, plasticity essentially posits that our brains are capable of changing shape and structure from learning or experiencing something new.

In 2009, Haier was commissioned by the makers of Tetris to follow up on his 1991 work with a group of teenagers, using much more advanced scanning techniques than were previously available. This time he found it much more difficult to find a group of people who had never played video games before. The new results showed that just as with the earlier research, the brain regions were heightened notably during initial game play. The study also, and more importantly, showed that just like the jugglers whose brains changed shape, the gray matter regions of the brain grew while the participants learned to play Tetris.

Bang, Bang!

Several times a week, a friend and I click into our Xbox for a little game play. In a matter of moments, we are knee-deep in a brutal war, taking on roles as commandos, privates, and sergeants and handling guns and grenades. Quickly, I'm completely engaged by my imaginary job of protecting my country, working with and competing with my buddy as we shoot and destroy enemies and strategize through realistic video images.

I genuinely enjoy this. It relaxes both me and my friend after a long day at work. Video games are, to me, an especially compelling form of storytelling because they allow me to control where the story goes and dive directly into the narrative, using joysticks and buttons. I find my game, Modern Warfare 2, fun and rewarding, and it refreshes me for the work I have to do.

One reason it may be so pleasurable is that like many experiences that are rewarding or exciting, playing games may stimulate brain dopamine, a chemical that plays a part in motion, intelligence, and especially pleasure. Steven Johnson, who has written several books on technology, including *Everything Bad Is Good for You*, argues that television, video games, and other "bad" forms of entertainment are actually good for our brains and creativity. He has written that the neurotransmitter dopamine is constantly being stimulated when we're playing games and is essentially responsible "for both reward and exploration." The stuff, he adds, is "the brain's 'seeking' circuitry, which propels us to explore new avenues for reward in our environment."

But war games like Modern Warfare 2, which put people like me into a lifelike role as the shooter, have, like porn, come under a fair bit of fire from people who fear that the games distort players' perceptions of reality and lead them to become comfortable with gratuitous violence. Admittedly, there are some issues with the violence in some games and there are concerns about kids who are so addicted to playing that they cannot stop. But these kinds of games, it turns out, also have some profound positive effects on the brain and our abilities.

Neuroscientists first began looking at the effects of video games on brains in the early 1980s as games such as Pac-Man and Donkey Kong became worldwide phenomena. Research showed increased visual skills and better hand-eye coordination. One study in 1989 tested hand-eye reaction time by asking people to press a button when they saw a light. The participants were then split in two groups and asked to play an Atari game system for fifteen minutes. When they were tested again, the game-playing group increased hand-eye coordination almost 50 percent. That's pretty powerful learning.

Then there was Haier's Tetris research, along with other game-related findings through the early 1990s. But a big break in the power of video games and neuroscience was actually discovered by accident.

Daphne Bavelier is the director of the Brain and Vision Laboratory at the University of Rochester, and although her career didn't start out this way, she now studies the effects of video games on vision and spatial awareness, specifically the impact video games have on cognition, brain plasticity, and vision.

In 2003, Bavelier and a researcher began to look into learning and brain plasticity and how new kinds of visual stimulation can affect the deaf. One of Bavelier's PhD candidates, Shawn Green, was preparing to test a visual computer system on a group of deaf participants. Before his formal tests began, Green tried out the test to make sure the instruments and data collections were all working correctly. This specific study was intended to measure an individual's visual acuity by identifying a series of dots on a screen.

When Green took the test several times to make sure all the instruments were working correctly, he saw that he was consistently getting perfect scores on the visual attention portion of the test. Assuming there was a bug in the program, Green asked some friends to come to the lab to take the test too. Green and Bavelier soon found that some people consistently scored dramatically higher than others did. After investigation, the research team discovered that those with nearly perfect visual test scores had one commonality: They consistently played video games—first-person shooters, to be precise.

Bavelier ended up studying the players of shooter games and had remarkable results: As a group, these players were not just faster in various hand-eye tasks, they seemed to have greater brain capacity, seeing more with peripheral vision, switching attention from one thing to another, tracking multiple items, and generally showing superior visual skills. These games, which demanded quick reflexes and accuracy, were more effective than games of strategy or role playing.

The research created a bit of a firestorm after a story about it appeared in the *New York Times*, with the startling headline "Video-Game Killing Builds Visual Skills, Researchers Report."

Sadly, the focus of the research was lost amid the outcry over the apparent support of first-person shooter games. If people were able to put the content aspect of the argument aside, they would see that Bavelier's study indicates that game playing clearly has a positive side: The skills that allow game players to move quickly, aim, and make superfast decisions can be transferred to a completely different kind of task. But many people tend to overlook the positive side of the research on video games because they have preconceived notions. Bavelier explained in an interview how frustrated she was that people didn't see the positive side of her research because they couldn't see past the fact that first-person shooter games were used in the studies.

Green and Bavelier's research over the last five years shows that action-video-game players consistently outperformed people who didn't play video games in multiple visual and hand-eye coordination tests. Their research shows action-video-game players have better "spatial distribution and resolution of visual attention," more "efficiency of visual attention over time," and a greater "number of objects that can be attended to simultaneously." Although the practical applications of these skills will vary on an individual basis, this can translate into being a better driver, a more practiced pilot, or a more accurate surgeon or even being better at navigating the Web.

Although it is up to each individual to find a balance with game play, these findings argue for more game playing, not banning kids from playing, and more interactive and active opportunities. Already, new games and game consoles such as the Nintendo Wii allow players to literally swing tennis rackets, dance, do exercises, and participate in other physical activities while playing. Microsoft's Project Natal creates an augmented reality gaming experience in which you become the actual game controller and there are no buttons or joysticks to worry about. You can play the game by standing in front of your TV and kicking your legs in the air, thereby kicking a ball on the screen. Mobile augmented reality games encourage players to go outside and run around by chasing a figment of a digital reality on mobile devices, blurring the line between sports and video games. This type of game play should be encouraged and supported, not ignored simply because the words "video" and "game" are in the same sentence.

The fact that these kinds of games are coming along is a good thing, since the genie is already out of the bottle and he's too big to be stuffed back in. An estimated 97 percent of kids twelve to seventeen years old play video games, and their entertainment is often not solo. A Pew Research survey asked kids how they play games, and although some like to play alone, 27 percent said they play with a friend online and 65 percent said they play with a friend or group in the same room—a different look but the same experience as a rousing board game of Monopoly or War.

On top of the social aspect, most of the games young people play are fairly tame. In 2008, when the Pew researchers asked the participants to list the top ten games they play on a regular basis, only three games turned out to be first-person shooters. The rest included solitaire, Tetris, racing games such as Mario Kart, and numerous sports games. Of the 2,618 games mentioned, the number one game among kids was Guitar Hero, a game that requires multiple players to get up off the couch and compete by playing a guitar and drums as if they were actually part of a band even if they've never had a music lesson in their life.

These players aren't likely to give up their games any more than I am. And like me, they most likely will play games as much as they read, testing and expanding their brains in different ways. Video games offer engaging, immersive, truly multimedia storytelling and can draw in participants more powerfully than can many traditional storytelling methods. That said, they don't replace one medium with another. Instead, they fill a new void created by a need for interactive narratives.

It's important to note that there's a place for each medium. Video games partially displace some forms of storytelling and in other instances meld to form new scenarios. Reading, for example, drives creativity in the brain in ways that video games can't. A careful selection of words can help our minds imagine, visualize, and daydream. Good written narratives offer a captivating path to the imagination and are imperative to our comprehension and reason. Stories that are told through audio help our brains learn to imagine in other ways and perfect our auditory senses. Images and video offer skills in visual perception and objective thinking and a different kind of logic. Video games offer a challenge to our cognition, coordination, working memory, and visual engagement, among other areas of the brain.

All these media engage our brains with equal weight and importance. The Web offers a culmination of everything for our brains through a new form of narrative and engagement that pulls us, along with our brains, into a new era of storytelling.

6

me in the middle
the rise of me economics

"I thought you were going to read the news," I said. "This is my news," she replied.

The New You, Always in the Center

If you pull out your smart phone and click the button that says "locate me" on your Google or Yahoo! map application, you will see a small dot appear in the middle of your screen.

That's you!

If you start walking down the street in any direction, the whole screen will move right along with you, no matter where you go. This is a dramatic change from the print-on-paper world, where maps and locations are based around places and landmarks, not on you or your location. People don't go to the store and say, "Oh, excuse me, can I buy a map of me?" They go to the store and ask for a map of New York, or Amsterdam, or the subway system. You and I aren't anywhere to be seen on these maps. The maps are locations that we fit into.

But today's digital world has changed that. Kevin Slavin, a creator of location-based services and games and the co-founder of the gaming company Area/Code, put this succinctly at a technology conference last year: "We are always in the center of the map."

Though Slavin was talking about location-based games and Google maps, the center of the map, it turns out, is actually much bigger than a dot on the screen. It's a very powerful place to be.

Being in the center—instead of somewhere off to the side or off the page altogether—changes everything. It changes your conception of space, time, and location. It changes your sense of place and community. It changes the way you view the information, news, and data coming in over your computer and your phone. And it changes your role in a transaction, em-powering you to decide quite specifically what content to buy and how to buy and use it rather than simply accepting the tra-ditional material that companies have packaged on your behalf.

Now you are the starting point. Now the digital world fol-lows you, not the other way around.

This transition has been coming at us in fits and starts for some time. As a thirteen-year-old with goofy milk-bottle glasses, at a time when the Internet existed only through slow dial-up modems, I couldn't wait to get online. At the time, I had moved with my father from England to Florida, and my transition to America and teenagerhood wasn't going very well. My father, who had an engineering background, connected the computer in his home office to the Internet and signed up for

America Online's $19.95-a-month service. For that amount, AOL rationed minutes of connection time like they were gold, allowing us only double-digit minutes of online time a week. This sounds ludicrous today, when we have unlimited Internet service for $25 or $30 a month, but in the mid-1990s those minutes were worth every penny.

When I walked in the door from school, I would beg to dial in, even if just for a minute. Going online was completely different from anything I'd ever done before. I could connect and chat with teenagers on the other side of the planet. The fact that I was "talking" to another thirteen-year-old in China or France was undeniably magical. It opened my eyes to a world outside the ten houses on my cul-de-sac.

I could search for answers to questions I had in my homework assignments by using some crude "interactive" encyclopedias or even ask online strangers for help. I'd feel a quick pulse of excitement when I heard the speakers on the desk (in a monotone computerized voice) say, "You've got mail!" But best of all, I was in the driver's seat, completely in control of where I went and when. There was no predetermined beginning or end. Even in the embryonic years, I was at the center of a World Wide Web experience

As we've seen, the power to be in control expanded to other content areas. Putting the viewer at the center forced porn to move beyond the blond-haired, blue-eyed beauty to address all kinds of tastes and offer content tailored to an individual's personal interests. First-person video games allow you to navigate on your own terms, land the plane yourself, or become the warrior or the alien. It's the difference between playing a car-racing game in which you're watching from the sidelines and playing one where you are sitting in the driver's seat with your hands on the wheel. You are part of the story, not just an observer who watches and cheers.

The twenty-first-century creation of elaborate social networks even more emphatically puts consumers in the center of their own complex web of connections and anchoring communities, those crucial networks that help make sense of the vastness of the Internet. If I go online today, I'm better connected with people from other countries than with people who live in the same city. I actually don't know where half the people I interact with online live, and it doesn't really matter. My only concern is their relevance to me, and vice versa.

This same relocation, this same centering of each of you in the middle of your own map, is also changing the concept of media. The word "media" has its roots in the word "medium," the middle man, and that's the role the media played, providing art lovers access to artists, readers access to writers, and citizens access to news.

But nowadays, if you're a media company, you might as well break the "dia" from the end of the word. As far as the modern young consumer is concerned, when it comes to content, there is only "me." Today. Right now.

I got my own hard lesson in this new Me! Now! world when some friends stopped by our house with their teenage cousin Lauren. As I started making coffee for our guests, Lauren asked if she could use my laptop to "check the news." I handed it over.

I was curious about which news sites she was going to, so I asked her, expecting to hear something like CNN or NYTimes, or maybe TMZ, the Hollywood gossip site. With a sincere face she looked up at me and said, "Facebook." Then she turned back to the computer and continued reading.

"I thought you were going to read the news," I said.

"This is my news," she replied. *News for social — the choice, not controlled by media like newspaper, TV, magazine etc.*

To Lauren and many in her age group, news is not defined by newspapers, or broadcast television stations, or even bloggers or renegades. Instead, news is what is relevant to the individual—in her case, what Facebook calls its "News Feed." The feed "highlights what's happening in your social circles" and produces "the latest headline generated by the activity of your friends and social groups," the company explained when it introduced the novel service in September 2006. And although users initially recoiled at the notion of so much personal detail being shared, the news feed has become a crucial part of Facebook's service and was, in some ways, a precursor of today's Twitter. Lauren and others like her are still consumnivores, gobbling up many kinds of content—but they're very particular and demanding about what they will eat.

When Lauren and her friends sit down at their computers and go to Facebook.com in a Web browser, they truly believe they are reading the news—*their* news. Though she may see "news" differently than I do, I am doing much the same thing when I go through my Twitter feed at night and in the morning and use that as my own highly personalized "newspaper."

In all the hand-wringing over the revolution roiling traditional media, this shift has largely been ignored. But it's central to understanding what has changed and what the future might look like. In the aftermath of the gory and devastating World War I, the poet William Butler Yeats wrote,

Things fall apart; the centre cannot hold;
Mere anarchy is loosed upon the world.

A couple of generations later, the writer Joan Didion surveyed the social revolution of the 1960s and wrote, "The center was not holding." In the middle of a painful technology and information revolution, the publishers, producers, and purveyors of traditional media may well feel the same way, that the center has dropped out altogether and a kind of digital anarchy reigns. It's an understandable response. The digital anarchy we're experiencing today has torn apart markets as they have been known for hundreds of years, replacing them with something still taking shape and yet to be determined.

From my perspective, though, the center of the media world isn't gone. It has drastically shifted in an earthquake-like moment. The birth of the Internet was the beginning, but we will feel the aftershocks and tremors for years as we move from a broad audience of readers or viewers to a very narrow audience of me and you, each a target market and each always in the center of the map.

A Greedy Audience of One

As core as this concept of a new center is, I am struck by how difficult it has been for traditional media companies to recognize it and address it head on—though their struggle is completely understandable. Their business models for generations were built around the idea of delivering a specially selected mix of content to a broad audience. The sales pitch of the music industry wasn't "Here's a good song you might like." It was "Here are a couple of good songs you might like and ten or eleven others you probably won't like—but we need to fill a CD so we can justify selling it to you for $15!" But for the past several years, sales of CDs have fallen as much as 25 percent a year and dollars sales of digital tunes haven't grown nearly enough to make up for the decline, according to the Recording Industry Association of America. *draws faster* -

For magazines and newspapers, the equation worked almost the same way. A thick stew of general content drew subscribers and readers, which drew advertising. But paid circulation has been sliding, and in recent years newspaper print advertising revenue has dropped sharply. What has happened is no secret. You'd have to have been living on the ocean floor over the last decade not to know that more and more content has been available for free, some of it offered by the companies themselves, such as online newspapers and blogs, and some pirated by enterprising consumers eager for the latest music, movies, and headlines. Yet even as one content industry after another has been hammered by new media stealing eyeballs from old media, some executives have resisted offering consumers a choice of how they would like to consume their content.

The music industry, for instance, took a walloping in the early 1990s but finally got back on its feet by agreeing to sell songs for 99 cents in the iTunes store. Yet recently, the record labels decided to start charging $1.29 per track for "hit new songs" because there is more demand for them. To many consumers, this comes across as greedy and unnecessary. When Sony announced its new digital Reader in 2009, the company said it had struck a deal to distribute software that acts like a library, where individuals can rent digital books on their eReaders. But there was a catch: Publishers provide a limited number of "licenses" for each book, so if someone else has "checked out" a digital version of Cormac McCarthy's novel *The Road*, you have to wait until that person "returns it" before you can download a copy. Consumers who can easily share a photo, song, or article with thousands of people and aren't tied to the limitations of a physical object have a hard time understanding this—and none of these transitional solutions make sense or adapt to a me-filled world.

Take the missing "me" in the movie industry. Although I still love the long format and immersive storytelling of movies, I am increasingly weary of watching them at movie theaters. I don't want to have to be at a theater on somebody else's schedule, chow down on their overpriced popcorn, or risk having a Chatty Cathy behind me. Instead, I'd much rather watch a movie at home, starting whenever I feel like it, with a bag of microwave popcorn, free tap water, and a handy little pause button for those inevitable restroom breaks.

A lot of people won't agree with me. The 3-D experience and digital technology helped bring in record box-office revenue in 2009. But why not give us a choice? The industry insists on holding back digital downloads and DVDs for some months after a film is released rather than giving consumers a variety of options for viewing the latest films. Yet even with that strategy, sales of DVDs fell sharply in 2009 as viewers found more affordable—or free—alternatives for watching movies from their favorite chairs.

Although the strategy seems to be working in favor of box-office sales at the moment, it may be just a momentary victory before things fall apart. We've seen this happen with just about every other industry, starting with music. People will find a way to get the content in the format, shape, and size that fit their desires.

A site called Pirate Bay that offers links to download movies, music, and books has 4.5 million visitors every single day. These visitors click on more than 26 million pages on the site, downloading whatever meets their fancy: *Iron Man 2*, an episode of *Family Guy, Frank Sinatra's Greatest Hits*. And this is just one site. A site called Torrentz.com, a similar file-sharing home page, brings in 2.6 million visitors a day and delivers nearly 14 million page views in the same twenty-four-hour period. On top of this, there are literally hundreds of these sites all over the world serving tens of millions of users who want their movies and want them now.

Now, it's true that some people steal movies, TV shows, e-books, and other digital paraphernalia just because they're there, but many people steal them because they're not there, at least not offered by the people who create and sell them.

I have to admit that I'm guilty of this behavior myself. In November 2007, two weeks before *American Gangster* was set to open in movie theaters, a copy of the film leaked to the Internet. Digital content can proliferate online faster than any virus on earth, and I decided to download a copy. It wasn't that I wanted to save $10; had I been able to buy an online version, I would have done so without blinking an eye. I simply wanted the experience of watching it at home.

Ironically, a few months later, I met one of the executive producers of *American Gangster*. Sheepishly, I shared my sin of "illegally" downloading his movie. I even tried to give him

$10 for the cost of seeing the movie, which he graciously declined to accept (although I think he really wanted to take it).

When I asked why he wouldn't offer the movie as a pay-per-download or something similar right from the start, he had two responses. First, he said, his industry "can stop the pirates and halt illegal downloads."

My response: "You're never going to stop an eighteen-year-old kid in his bedroom in Sweden with too much time on his hands and a zeal to do something he's been told he can't do from finding a way to put your movie online." I tried to remind him about the music industry and its failed attempt to stop music sharing as individuals created their own personalized and specialized mixes, but he only scoffed and told me that his industry is "smarter and has deeper pockets."

OK, enjoy emptying those deep pockets, I thought.

Then he explained that people prefer the movie theater because it represents a campfire type of storytelling that humans have shared for thousands of years. That is, we like to get together to have shared experiences, to listen to a compelling narrative. Perhaps, I thought, but my campfire is now digital and easily shared. We don't necessarily need a communal theater when we can sit around our own television screens and exchange comments with our friends in a digital context.

Just like every transition taking place, this isn't a black or white answer, and I don't mean to sound like the wags from the *New York Times* way back in the 1880s who worried that having phonographs would mean that people would never attend a concert again. There are times when a movie theater is the right fit for me and my friends. In some instances, it's nice to get out of the house and drop into a large theater to laugh with friends at a comedy or experience an intense action-filled movie on the big screen. But more often, I prefer my home theater with a personalized start time and that important pause button.

As for the "campfire mentality," I agree that humans love to sit around a real or figurative campfire. But a campfire can happen in bits, too. I often message friends asking what they thought of a movie before or after I see it or keep an eye out on other social sites to see what the campfire sitters are saying. And I also add my views.

A movie's digital campfire was demonstrated in March 2010 when Sitaram Asur and Bernardo Huberman, two researchers from Hewlett-Packard's Social Computing Lab, used Twitter to predict movie box-office sales by monitoring the commentary and opinions of strangers on the site. Asur and Huberman monitored almost 3 million tweets to predict if people thought a film was good, bad, or indifferent. From that, they predicted the success of a new movie in the theater.

How did they do? They found that people sharing their views of a new movie on Twitter could foretell with 97.3 percent accuracy how well, or poorly, a movie would perform in an opening weekend at the box office.

What the researchers didn't know was how many people were actually watching the movie at home, probably illegally, or around a "campfire" at a theater. That's where the website torrentfreak.com comes in. TorrentFreak is a blog devoted solely to the goings-on of the file-sharing communities on the Web, including a well-known protocol called BitTorrent that monitors and reports download numbers and news related to policy and law around person-to-person file sharing. Every year around the Oscars, TorrentFreak's editors release the top ten downloaded movies of the year, a tradition appropriately called the BitTorrent's Oscar. The number one downloaded movie for 2009 was *District 9* with 12.6 million downloads. Second on the list was *Avatar* with 11.3 million downloads. This doesn't include pass-along rates, where people share the files with their friends. These movies aren't just being downloaded by a few kids in their bedrooms; they are being downloaded by tens of millions of people all over the globe.

The availability of music, words, and movies in a variety of different formats means I can personalize and customize my use of them. Impatient and demanding consumers increasingly won't want to wait for digital formats, and I believe movie distributors and others are missing an opportunity (and perhaps encouraging piracy) by declining to make varied formats available much more quickly and at a fair price.

Consider what's going on in the book industry. In early 2010, some publishers, including Simon & Schuster and Hachette Book Group, said they would delay making e-reader versions of many books available because they were afraid that electronic copies of the books would cannibalize sales of more expensive hardcover editions.

Carolyn Reidy, chief executive officer of Simon & Schuster, told the Associated Press in an interview, "We believe that a large portion of the people who have bought e-readers are from the most devoted reading population. And if they like the e-readers, they are naturally going to convert because the e-books are so significantly less expensive."

Well, part of that is true: Devoted readers bought e-readers because they want to read books on them. But to assume they are being cheap seems misguided. I own an Amazon Kindle, a Sony Reader, and an Apple iPad, but I didn't buy them to save money, nor did the avid bookworms I know who also bought one or more of these devices. How can e-books save money when a person spends up to $500 on a device and pays $10 or more for each book?

These are book lovers, right? They want to bring their collections with them without the physical weight. They enjoy extra functions such as being able to look up words in a built-in dictionary, sharing content with others, and taking notes on what they're reading. Most important, e-reader users want instant access to books in the airport, the subway, or a coffee shop. Once they are intrigued by a new book, they can start reading it a couple of minutes later. In fact, it seems to me that e-readers could increase book sales by making books more accessible than ever. (A 2010 survey by L.E.K. Consulting, a business and strategy consulting agency, found that 48 percent of e-reader owners said they were reading more books than ever, compared with 7 percent who said they were reading fewer.)

Understandably, publishers are worried about changing business models and what will happen if they charge less for electronic books. And the $10 price that consumers expect, originally set by Amazon.com to build market share for the Kindle, may well be forcing them to sell below cost, never a winning formula in any business. But do they really believe they will boost their bottom lines by simply trying to keep devoted readers away from digital books? No. The digital torrent sites I referred to earlier don't just share movies and music— they share e-books, too.

I wrote about this for the *Times*, saying, "Let's say you un-wrap your birthday present and see the latest Kindle, Sony Reader or Barnes & Noble Nook. Just what you've always wanted! You turn on your new device, navigate to a wireless bookstore and search for Don DeLillo's new novel. Instead of a simple click and download right from your armchair, you're told it's available only in hardcover for the next four months. Are you really going to get in your car, drive to the store and buy the hardcover?" Instead, you're likely to buy something else from the digital bookstore.

(Can you imagine if the digital camera you just purchased gave you this warning: "We're sorry. You won't be able to e-mail this photo to your friend for another four months. Instead, why don't you print a copy and mail it through our on-demand printing service"? It's hard to imagine that any buyer would be happy with that.)

These publishers seem to be picking a fight with the wrong team: their customers. They are punishing the people who buy their content instead of making it simple for those customers to hand over their money instantly from any location in the world.

That said, only a few publishers are joining this process. Most of the other publishers I spoke with when reporting the story for the *Times* said they would continue to release books in print and in digital at the same time, and that's a smart move, given how quickly e-books are gaining ground. Jeffrey Bezos, Amazon's chief executive officer, said in 2010 that if Amazon has a Kindle option available to readers, it will sell forty-eight Kindle versions for every hundred copies of the physical book. "It won't be too long before we're selling more electronic books than we are physical books," he predicted.

With e-readers proliferating, it also won't be long before we have the option to read or watch just about anything—magazines, newspapers, movies, TV shows, or church newsletters—on a portable reader. A generation already in its teens probably will mature and enter the workforce believing that all of its media bytes, snacks, and meals will be delivered on a screen. The limits of paper won't exist. Digital will mean "immediate" and "infinite" and "extremely personalized" for the customer at the center of the map.

"Me" Economics

"OK, great," you say. "So we're moving to this world of digital narcissism where consumers young and not so young aren't just always on their phones or texting away but also demanding that they have their own customized and personalized music mixes, movie choices, and specifically culled news selections. Who is going to pay for this great music, these fabulous movies, and these very important (and expensive-to-produce) news stories?"

Great question! As an employee of this industry, I've been involved in more meetings on this topic than a single human being should be allowed to attend in a lifetime. I've been to all-day talks and fifty-person meetings with everyone from the CEO to a lowly intern and all the players in between. I've also attended conferences as a panel member with other journalists and publishers to discuss this very topic. Depending on who is in the room, these conversations usually start out lamenting the death of newspapers and magazines and quickly move to asking how we will ever be able to charge people to access news online.

I have heard over and over that young people won't pay for anything. Movie producers, publishers, and musicians argue that kids have been raised to think content is free and they have a God-given right to take it.

I'm not going to list a magic formula here with profit margins, returns on investment, or revenue models. That is simply not my area of expertise. But as a student of new technology and consumers, I use a four-prong formula when deciding whether to purchase digital content: price, quality, timeliness, experience.

- People will pay for some experiences around the content but not for just the content. But people *will* pay.
- They will pay for quality, whether it's high-level graphics, a beautiful design, or graceful language.
- They will pay for timeliness if the experience of having something first or before it perishes is worth paying for—if they can purchase it immediately.
- They will pay if the price matches the experience. Just as with porn subscriptions, in which the sales drop off once the price hits a certain point, there will be a limit to what people will pay for content. The amount may be below the seller's hopes—but there is a price people will pay.

I still pay for digital content all the time. I buy the *New Yorker* on my digital e-reader. I buy heaps of fun applications for my phone and games for my Xbox; sometimes I even buy TV shows and music for my iPad. What makes me decide when I will pay for music or TV shows or films? In short, I base my choice on the overall experience and what I want at that particular time. Here are three different ways people, especially young ones, may evaluate whether something is worth purchasing.

Bad = Free

*if he really love it.
he will buy it.*

My friend Mike loves music. In fact, Mike is a music fanatic. In every spare moment he has, Mike scours the Web and his social networks, searching for new music to listen to and potentially purchase. Like most of his friends, Mike uses his recommendation systems and social networks to find the music he's interested in. He'll preview a few songs, and if he decides the content is good, he'll follow through with a purchase. He rarely buys entire albums because he believes most albums contain only one or two good songs. Mike also follows a handful of bands and immediately buys their entire albums on release day.

But Mike steals music, too. He doesn't steal music because he can't afford it or to take a stand against media moguls and corporations, and he definitely doesn't do it for the thrill. He does it for two simple reasons: Either he thinks the content is overpriced or he wants payback for something he purchased that was unsatisfactory. That is, sometimes he'll buy a couple of songs and then download others, figuring the total he paid balances out to a fair amount.

If he previously purchased an entire album and felt that the majority of the album was inadequate, he feels cheated because there's no way to return the album. The next time, he'll find that artist's work for free online and download it, essentially stealing it.

You might think this is a lazy and ridiculous justification—illegal, plain wrong, or perhaps a sign that civilization as we know it is ending. But Mike is frustrated that poor-quality digital merchandise or disappointing downloads can't be returned like a shirt that doesn't fit properly or doesn't match other clothes. He knows that free music abounds on the Web for those tenacious enough to find it—much like those of us who bypass the dealer to buy "used" hubcaps from the resale shop even though we know we may be buying stolen merchandise. Mike has established his own Internet law of economics and can implement it because so much music is easily available for free on the Web. To him, it all balances out.

You might think Mike is one in a million, but he's not. I've heard many people say they do the same thing. An individual who works in politics has the same mentality. When I asked why he steals music or if he feels guilty doing it, his response was an abrupt "absolutely no way. I feel deceived if I buy a whole album and 90 percent of the music is bad."

Peter Serafinowicz, a British producer and actor who has appeared in more than forty TV shows and movies, including *Shaun of the Dead, Star Wars,* and *Couples Retreat,* admitted to piracy in May 2010 in a blog post on the technology website Gizmodo.com. In a piece titled "Why I Steal Movies . . . Even Ones I'm In," he said that a main reason he will download illegal content is that it's not available for sale on the Web. So he'll hop online, do a quick search, and find the show or movies he wants, which usually arrive on his computer in moments.

Serafinowicz wrote that he hopes people will steal his own TV show. He explained that when his new BBC TV program *The Peter Serafinowicz Show* appeared on illegal file-sharing sites on the Web, he saw the posts as a way to spread the word about his new, relatively unknown show. In fact, he added that he's illegally downloaded the show himself, since it's easier than trying to find a legal copy online.

Serafinowicz said that he'll pay for a show or movie—but only when it is "better than free." "I'll click and buy," he wrote. "It's simple, quick, better quality, not to mention legal. It's also cheap."

Serafinowicz explained that he practices his own version of Me Economics. "I recently wanted to show my son Disney's classic *Jungle Book* and intended to get it on iTunes," he wrote. "Unfortunately, it is currently incarcerated within the Disney Vault. So I'm afraid I simply downloaded a pixel-clear pirate copy which arrived in seconds. My moral justification for this? I once bought the VHS. It's your own vault, Disney!"

Not surprisingly, many commentators agreed with him. One wrote, "Lesson to content providers: You make it easy to own, or we'll make it easy to own."

In other words, when consumers are not offered an option, they make one themselves.

Price Is Relative to Cost

Consumers of online and digital content are all too aware that what they are buying in many cases costs far less to produce than the old-fashioned product. Creating a digital version of content costs the same whether you're creating 1 copy or 10 million copies. It costs practically nothing to re-create bits, beyond hard drive space. So consumers expect costs to change accordingly for the digital version.

Newspapers get to drop the paper, printing presses, and delivery. Books don't have to be transported or stored. Music doesn't have to be pressed onto CDs and shipped to stores. Anyone with a Flip camera can make a video; anyone with a digital camera can snap a shot of a fire or tornado or another news event and share it with the local newspaper or television station. Any wishful novelist can self-publish a book that looks an awful lot like what you see at the store—even if it doesn't read the same.

Bill Grueskin, dean of academic affairs at the Columbia University Journalism School and former managing editor of WSJ Online, noted that the cost of new subscribers fell once the business was established. The *Wall Street Journal* originally sold a subscription to its website for $49 a year. But "once it had established a base, the incremental cost to serve a new subscriber was $8. By 2006, the cost of the annual subscription had climbed to $99—but the price of serving that additional subscriber was a mere 85 cents." Of course, without a print edition that is supported by print circulation and advertising, the costs to produce editorial content would be different. But these striking numbers underscore that it is less expensive to produce digital copies without the costs of printing, paper, and physical distribution.

Of course, there are still costs—highly skilled editors, copy editors, author royalties, and so on—but distribution is drastically less expensive, and the public recognizes this. In the public mind, the product you hold in your hand should cost more than the one that was downloaded—especially if it was downloaded onto an expensive e-reader or another gadget.

As technology has demolished barriers to entry, consumers have become much more aware of what it costs to produce new content. Now anyone sitting in her bedroom with a microphone and a laptop can become a music producer. You don't even need a separate camera and a tripod to create a TV show. Using only the camera built into the computer, young producers have made videos that collectively have reached hundreds of millions of viewers on YouTube and other advertising-based online video outlets. Mike Wesch, a YouTube anthropologist, says that a single famous video of a kid dancing in his bedroom to the Numa Numa song has been viewed more than 50 million times. And the cost to create, edit, and distribute this video was probably close to zero.

This became apparent in the music industry in 2007 when Kate Walsh, a solo guitarist from the United Kingdom, decided to record an album of her own music. She went to her friend Tim's house and spent a "few hundred pounds" (mostly on thick velvet fabric to soundproof Tim's bedroom) to record an album that she released digitally via iTunes. Before she knew it, she had the number one album on iTunes, quickly surpassing the extremely famous band Take That.

In an interview with the *Evening Standard*, a London newspaper, Walsh said, "I set up my own record label called Blueberry Pie and just got the music out there. It's pretty easy. Anyone can do it." When asked about the costs of recording and distributing her own album, she replied that "you don't need loads of money to make an album and they don't need the backing of a record label. There's no advertising or marketing involved, you don't go on how much money has been spent."

Though such fame is admittedly rare, Walsh isn't an outlier in the music world. A couple of years ago, Justin Bieber was just entering adolescence and living in low-income housing in Stratford, Ontario, when he uploaded a few videos of his singing to YouTube. By accident, a hip-hop marketer named Scooter Braun stumbled across his songs and then tracked young Bieber down. To build up his experience and his image, he flew Bieber to Atlanta not to make an album but to make more YouTube videos, filmed by other kids rather than using expensive equipment.

Hip-hop star Usher heard the songs and leaped at the chance to sign the mop-haired boy wonder. Before Bieber turned sixteen, he had made two albums, become the biggest teen sensation in a generation, gone on tour, sung for the president of the United States, and appeared at Madison Square Garden. He is cashing in, but his original work cost almost nothing to produce.

Now that we know this, why would we pay loads of money to buy an album unless we wanted it very badly? Understandably, content industries want to charge as much as the market will bear—but the market won't bear as much as it used to.

Price = Quality of Experience

If you offer unique quality content at a fair price, I guarantee people will pay for it. How can I guarantee that? Apple, the computer and music company, has already done the research for me.

Comment!

Before iTunes came along, my friends and I stole music all the time. Sure, you could buy songs and albums online, but the choices and quality were extremely limited, and it's an understatement to say that the process of actually buying the music was painful. Downloads were slow, digital music was extremely limited, and getting the music onto a digital device required a computer engineering degree and a lot of patience.

So we started to steal all of our music, all the time, years before social peer-to-peer sites such as Napster and TekNap arrived on the music pirating scene. Instead we would go to hacker sites called trackers and search for and share the MP3 files and albums we wanted. These peer applications and trackers, unlike legitimate digital music stores, were incredibly simple to use.

In 1998 I remember buying one of the first consumer digital music players, the Rio PMP300. The Rio looked like a regular Walkman without the tape slot and stored a whopping ten songs. I remember how excited I was to get the player, and as soon as I picked up my new fancy gadget, I was ridiculed by my friends for spending $200 on a music player that could barely hold enough music to listen to an entire CD. Even the salesman at the electronics store looked at me like I was crazy. I remember him shaking his head, saying, "These digital music players are a fad. You should just buy a CD player instead. It's much cheaper."

Turns out, they were right in ridiculing me. Buying music online and getting it on the Rio took longer than driving to the store and actually buying a physical tape or CD. Transferring music from my computer took twenty minutes. The user experience was utterly terrible, and the price was ridiculous.

Digital music players and online music stores promised a great experience, but the technology just wasn't ready for prime time. So despite my new expensive digital music player, I continued to steal music.

My music heist came to a screeching halt in 2003, a couple of years after Apple rolled out the iPod and opened the music store iTunes. MP3 players had come a long way since my Rio, and there were plenty of choices. But Apple had introduced coolness, immediacy, and simplicity; with the single click of a button, I could download, transfer, and listen to an entire album. From a click of the mouse to the play button on my iPod, the entire transaction took less than thirty seconds. And since the only way to do this was to buy the music, I happily bought it.

I clearly have lots of company. Since the iTunes Store opened for business seven years ago, consumers have downloaded 10 billion songs. Granted, not all of these songs are purchased—some are offered free, and others are included in longer albums as add-ons. But even if they gave away a generous 25 percent of these songs, consumers would have paid for 7.5 billion digital songs. That's a lot of music and a lot of money.

iTunes displaced a large proportion of music theft because it was simple, ultrafast, consistent, and one of the few ways to get music on your iPod, which quickly became a status symbol. The standard fee that Apple charged seemed reasonable and sensible: 99 cents for a single song and $9.99 for a whole album, regardless of the band's name or status.

That's part of the balance to be had with Me Economics.

Finding the iTunes Formula

The iPod and iTunes proved that we'll pay if the price is right and the experience is special enough. The same can be applied to other kinds of media. Let's say I subscribe to the print *New York Times* and have it delivered to my doorstep every morning. I pay $700 or more a year for this bundle of paper with black ink and pictures, but what exactly am I paying for?

I'm paying for the words, reported and written by some of the best in the business, naturally, but that's not all. I'm also paying for consistency, trust, good design, and an attractive layout. I'm paying for professionally produced pictures and graphics. I'm paying for the guy who wakes up at four a.m. and drives the truck to my house to deliver the newspaper to my doorstep. I'm paying for reliability, and I'm even paying for a little blue bag they put my paper in to protect it from the rain. I'm also buying the ability to discuss that article with my spouse or friends. In sum, I'm buying an informative and social experience.

On the Web, most of those services go away. I own the computer or phone where the content appears. If I use another news service and flow in a feed from the *Times*, I own the layout and design, too. If I were to pay (and the *Wall Street Journal* is just about the only outlet successfully charging right now), I'd seemingly be paying mostly for words. It would be for great content, to be sure, but in a world with commodity news (that is, content that's available just about everywhere) and a fair bit of useful free information, it would be a difficult bill to swallow—especially now that I'm used to having it for free.

My perception is that I'm not getting much of a special or different experience on the Web. I have to be at my computer, navigating takes time, all those links are somewhat overwhelming, and the content doesn't feel as personal or customized as the way I can work through the print edition. In newspapers and in other media, the packages haven't evolved much, yet the information is on a new platform and the experience hasn't really been transformed. It doesn't feel to me like something for which I should pay very much—or anything.

In reality, we don't pay for the content; we pay for the experience.

And there are digital experiences I would pay for. With news, for example, if I were offered a personalized and customized version of a digital newspaper that incorporated my personal preferences, location, and social circle or if the subscription software made reading especially easy, fast, and smooth, I would sign up right away. But right now, many online newspapers and magazines are only starting to swirl social customization, or me, into the experience.

A concert featuring one of my favorite bands is another example of paying for the experience over the content. I could easily buy an album for $10 or stream the music for free on the Web. But it's not just about the music—it's about the entire experience. People will pay—sometimes huge sums of money—to see and hear artists perform in the flesh, to hear the music as it is performed, to have social interaction, maybe to dance, and certainly to be entertained. You're paying for the entire experience, not just the music.

The same theory applies to books and other words on a page. Let's take the "reading on a screen" versus "reading on paper" debate out of the equation for a moment and look at the experience surrounding a book. Books offer content and information, but they are also comforting experiences. When you're reading, maybe you're lying on the beach with your feet in the sand, engrossed in a story. Or maybe you've curled up by the fire with chocolate chip cookies and a hot cup of coffee. Or maybe you're keeping yourself entertained on a plane. Are you buying just the words on the page? No. You're buying a cover design, the layout, and the opportunity to be enlightened. You're even buying the ability to discuss the book with your friends or coworkers or with a stranger at a cocktail party. Imagine if I said I would sell you this book on Post-it notes. Would you still want to read it? Probably not. The experience would be terrible to consume.

Taking this perhaps to the extreme, the words seem to be only a fraction of what you buy. Compare a hardcover book with the highly popular moleskin journals sold at the local bookstore. The book-sized versions of these journals cost $20—about the same as the price of many bestsellers—and the pages are blank. You aren't going to take this blank notebook home, drop into a comfy chair, and just stare at three hundred sheets of stark white paper. But you feel a connection with it, and that expensive moleskin cover will make you feel that whatever you write or draw in that blank book will be much more special.

This illustrates a key reason why selling content online has been so difficult for so many media companies: The experiences that original books, newspapers, and CDs provide haven't translated into something as meaningful in the digital realm for that me-in-the-center customer. Entertainment and content purveyors want the public to pay, but they have functionally stripped away most of the original experience that connects each individual to the product. Not surprisingly, you won't pay anywhere near the same price for it. That would be like going to a local restaurant and hearing the chef tell you that she will charge regular prices, but she needs to use your stove, pots, pans, spices, plates, and silverware. Oh, by the way, you will have to wash your own dishes, too.

The people who sell entertainment and words and information for a living need to understand that they are selling much more than that. They need to adapt to sell new digital experiences and give people incentives to buy the whole package, not just the words or sounds. They need to convince young people who have grown accustomed to getting so much for free that these new experiences are truly worth paying for.

We're selling to a new audience, and we need to talk to them differently.

I don't want to sound like an alarmist here, but the big fundamental changes are yet to come. Yes, over the last ten years our culture has started to see some very remarkable transitions take place. But over the next five years we're going to enter an even more extreme digital metamorphosis.

Although the Web is a little more than twenty years old, we still don't have any pure digital natives in the workforce—that is, people who have grown up online. (Even nerds like me are older than the online world we know today.) When that group comes of age, it won't remember going to the store to buy a book or having to rent a movie on a DVD. They won't understand what it's like to watch a TV at a certain time instead of replaying or downloading whatever they want at that moment.

Last year, in the research labs at the *Times*, a colleague was giving a tour of our offices to a friend who is an advertising executive. The executive had his three-year-old daughter with him. As she bounced around the office, touching and investi-gating everything in sight, my colleague asked if she knew what a newspaper was. The little girl paused as she examined an electronic gadget in her hands, looked up, and said, "I don't know what a newspaper is, but I know my daddy gets one on his phone."

To this little tyke, the concept of a thirty-minute TV show or a four-thousand-word magazine article won't exist. She'll consume in bytes, snacks, and meals on devices and screens we haven't even heard of yet.

This dynamic and prodigious group of consumers, cur-rently in middle school and high school (at best), will be your coworkers soon enough. They will bring to the office and the marketplace a set of ideals and preconceptions that are already drastically clashing with the current mentality and mind-set that we've been comfortable and familiar with for generations. If you also want them to be consumers of your print or film stories and your storytelling, you'll have to give them an expe-rience that is clearly worth paying for.

It's in Your Pocket

To get a handle on what that special kind of experience might look, feel, or sound like for those at the center of the map, don't look any further than your mobile phone.

As phones have gotten more sophisticated, offering quick and easy access to the Internet, your calendar and contact list, and all the gizmos and games you can dream up, they have become almost an extension of ourselves. People who swore by a paper calendar now can't function without a phone. People are buying fewer watches and bypassing alarm clocks because the phone keeps time for them and wakes them up. More than a few people watch TV shows, listen to music, and read on their tiny little phones. When you're in the center of the screen, the phone is also central to your life, work, and connection to friends, family, and coworkers.

And guess what? Although the price of technology generally has been going down, the amount people pay per month for phone service has been going up as they add minutes, texting fees, additional phones for the kids, and now data plans to their monthly bills.

People form incredibly strong bonds with their phones, so much so that even those who once swore that they would never read anything on a screen have slowly started to change some aspect of their reading habits. Even true believers in the print experience may see that a screen can provide just as strong an experience as a textured piece of paper.

Imagine you're in a coffee shop, a park, or your office and I ask you to hand your jacket to a perfect stranger. I then ask the stranger to inspect the jacket. As you sit watching him touch and explore it, the experience may feel a little strange, and you probably will feel a slight twinge of anxiety and maybe curiosity. The overall experience may make you a little uncomfortable but probably won't leave you excessively anxious (unless there is something you wouldn't want someone else to discover).

Now, imagine that I ask you to take out your mobile phone and hand it to that same perfect stranger. As he takes your phone in his hand, pressing buttons and touching the screen, you may feel apprehensive even if the person can't read your private messages or e-mails. I know I would. In fact, I know I have.

One reason we feel so connected to our phones is that we have them with us all the time. Our mobile phones are constantly just a reach away, connecting us to the fabric of the Internet. But more important, the deep connection with these devices comes from the association and bond they provide to the people we love, care about, and interact with on a daily basis. This device, a small chunk of metal and glass the size of a pack of cards, has become an extension of our relationships. Although the phones have not replaced the relationships, we can feel such an incredible bond with our mobiles that they can become a surrogate for those relationships.

How does that surrogate relationship work? Consider some early psychological research with monkeys. Obviously, although understanding the connection between people and their mobiles was not the goal of that research, it does help describe the interdependence and emotional connection we have to our mobile phones—and why we feel this way.

In the late 1950s, psychologists debated the importance of "love" in society. Some leading psychologists believed that love was not imperative for survival even though it could be an important factor in living a good life. Food and water are integral to life, they argued, but love is not vitally important, just an added bonus.

Others, however, believed love was in fact an imperative part of life and survival, on a par with food and water; without love, they believed, people might not survive and society could fail.

At the forefront of this argument was a University of Wisconsin professor, Harry Harlow. Harlow believed people could not survive without love, and if they did, their lives seldom would be happy and their bodies would age more quickly because of this missing link. After years of research with neonatal monkeys, Harlow published a paper in 1958 titled "The Nature of Love," providing evidence that love, or the connection to surrogate-like experiences, is in fact crucial to our survival.

Harlow followed sixty newborn monkeys. He removed the infants from their mothers a few hours after birth, and his lab assistants fed them through a bottle. The goal of the first experiments was to see how the monkeys grew up when raised without a mother. As Harlow suspected, when completely isolated, the monkeys did not do well. When isolated for long periods, he wrote, the monkeys went into "emotional shock" and in some instances refused to eat and died.

One curious part of the research, something Harlow didn't expect, was that baby monkeys showed a strong connection to the cloth pads that lined their cages. He wrote: "The infants clung to these pads and engaged in violent temper tantrums when the pads were removed and replaced for sanitary reasons."

This led his team to push the experiments further, and they began creating fake monkeys made of wire and cloth to see how the babies would interact with those surrogates. Then Harlow's team performed a number of experiments around the need for love with infant monkeys and a mother.

In one of Harlow's famous tests, the researchers created and placed two fake mothers, one made of wire and the other of cloth, in the cage with the newborns. The wire monkeys held a bottle of milk and helped feed the infants. The cloth monkeys couldn't hold the bottle, but they were comforting to touch. The researchers found that although the infant monkeys would take the bottles from the wire monkeys, they spent almost eighteen hours a day attached to the cloth mothers.

As Harlow explained, the researchers didn't expect such a dramatic attachment to the surrogates, but the theory led the way to more research in the area of love and survival.

Those findings also led psychologists to believe that connections to comforting objects can be as important as actual human physical contact. Just as the cloth monkeys became surrogates for the infants, our mobile phones become something like surrogates for our close relationships. As a result, we not only have a dependence on these mobiles but in some cases start developing an actual bond.

Even an older generation that didn't grow up with these gadgets relies on them. My mobile phone is one of my cornerstone connection points to the world around me. Imagine, then, the depth of connection the next generation will have. That connection begins at an early age, deepens when people are teens, and turns them into "mobile natives."

The mobile phone is also quickly becoming the first portable all-in-one device. We don't use it just to talk to friends and family: We use it to check the news; update our status on a social network; take pictures; read books, magazines, and blog posts; and then share the content accordingly. At face value, the mobile phone becomes a hub of information, but its role is much more intense than that of just another screen on which to read and consume information.

Researchers recently explored the phone as a way for parents and teens to feel a connection as teens become more independent and set out on their own paths. The psychologists found that as teenagers start to leave the house without their parents and start to engage with friends and discover their independence, both the teenagers and the guardians feel a sense of relief when they leave the nest with a mobile phone.

The researchers believe the mobile phone becomes a "transitional object," a psychological term originally applied to toddlers' teddy bears and blankets. Transitional objects create familiarity and comfort and also help develop connections and bonds. The authors also see the mobile phone as a strange object that crosses the line between a commercialized product and a childhood connection. It thereby becomes an important bond between parents and children.

Marshall McLuhan, the renowned media theorist who explained the cultural importance of television, believed the objects we surround ourselves with become an extension of ourselves. McLuhan said the car is an extension of our feet and that our clothes are an extension of our bodies. McLuhan also believed that media are an extension of our ability and need to communicate.

Given the extraordinary developments in what phones can do, it's possible that over the next five years the mobile phone will become the single most important device in our lives. These phones, our constant companions, connect us to any morsel of information and, most important, connect us to people. In turn the mobile phone becomes an extension of relationships. Although the mobile phone does not replace our bonds with people, it extends and perpetuates them. Paper, radio, television, and even the standard telephone all allowed conversation and communication, but our mobile devices are highly personalized and instantaneous.

In numerous interviews, university-based human-computer interaction specialists and media theorists unanimously agreed that these minicomputers in our pockets are changing the way we interact with people and content.

BJ Fogg, from Stanford University, explained that the mobile phone will replace so many objects in our lives that it will become the hub for everything we do. He said, "Just think about the connection we have with our phones now. We personalize them, plastering our photos on the home screens, changing the color of the fonts, and then we use them to send text messages to our friends and update our social networks. We are completely dependent on these phones and feel an extremely strong connection to them."

Another professor, Dan Siewiorek who is the director of the Human-Computer Interaction Institute at Buhl University, explained that our connection to the mobile phone has gone beyond basic phone calls and connecting with people; mobile phones are also an object to consume information, much as newspapers and magazines were in the past.

The phone is becoming the device we use to read news and check up on the things we find interesting. And because we use a single device for these activities, we are becoming increasingly reliant on it as a main connection point to the world around us.

When we talk about switching from print to pixels, from paper to screens, we tend to fall into theoretical discussions about the connection we have with paper, from the smell of the glue that binds the book to the rough texture of the book jacket. But for digital natives and many immigrants, screens are beginning to play a similar role. Their experiences and relationships with their phones are becoming more mean-ingful and more important. And as we will see in the next two chapters, it is precisely those kinds of significant and powerful individual experiences that will compete for our at-tention and drive the successful media and the technologies of the future.

7

warning: danger zone ahead
multiple multitasking multitaskers

The statement "one cannot do two tasks at once" depends on what is meant by "task."

—Donald Broadbent

Warning: Distraction Zone Ahead

It's clear that our brains interact in a new way when we're online. As the Semel Institute's research (see pages 140–144) shows, some areas of the brain can be stimulated differently when one is reading a book or static print narrative compared with the Web, which in contrast affords a type of multitasking storytelling. But this assumption brings with it a new set of warnings and conjecture: Some say the Internet and multitasking with media are making our brains into one big distraction zone incapable of handling complex ideas or long-form narrative. I don't agree.

I've heard similar comments since I was a little kid. Over

and over, they showed up on my report cards: "Nick doesn't pay attention in school." "Nick is too easily distracted." "Nick's mind wanders too much. He needs to focus." You might think, Hey, nice guy, but he can't get much done. Too bad—he seems to have a lot of potential.

My troubles focusing and the conclusion that I had a problem weren't just a childhood phase but a fact of my life. The label stuck all the way through middle school, high school, and college and still applies in my professional career. My mind still wanders, and I have trouble concentrating on one thing at a time.

You're probably thinking, Oh, that's easy: He's got attention deficit disorder.

Whether that's the case or not, if you give me a pile of random tasks to get through, I can happily execute my assignment and accomplish quite a lot. My style of operating is what I call ricochet working. And I'd argue that it isn't really a dysfunction, merely a different kind of functioning and one that we'll see more and more. The way my brain works is similar, I think, to the brain wiring of kids today who are growing up in the online world. They're "digital wanderers," popping between all different types of media, content, and experiences, and of course they are "easily distracted." They will also probably be successful ricochet workers.

Some minds work in a partial ricochet fashion because of the type of content they consume and the devices they use to consume it. Some of this is due to the way computers evolved.

In the early days of Apples, Dells, and IBM PCs, the computer took several minutes just to load, and then you could handle only one or two functions at a time. Each individual program took a while to open as well, and you probably could do only one thing at a time. As processors got smarter and faster, your computer began to "multitask," giving you the concept of windows—multiple functions churning away in various boxes on the screen. In the meantime, though, each of us also got better at doing a few different things at the same time.

Anyone who remembers the early days of the Web went through a similar experience as well; just connecting to the Internet took several minutes. There were passwords, strange fax machine–like noises and a few clicks of the mouse, and then interminable delays as the "World Wide Wait" slowly dripped into view. People kept themselves occupied by picking up a book or a magazine that sat close by, playing solitaire on the computer, or simply staring off into space, letting their minds wander.

Gradually, as computers became faster, they enabled us to perform multiple tasks simultaneously. Rather than wait two or three seconds for someone to reply to an instant message, I can read a few more words from the article in my browser or play another few seconds of that video game I started earlier. We have adapted to a world where information moves very quickly and in many different forms, from the television, to the radio, to the computer, to the mobile phone. As technologies change and we become more adept in using them, our brains will adapt too.

The Great Multitasking Debate

Whether this intensified jumping from task to task is a good thing is a subject of intense debate. It might make us smarter, faster, and more agile. Or it might, in the opinion of some researchers, simply make us more stupid and prone to destructive errors. We could become like the characters in Kurt Vonnegut's short story "Harrison Bergeron," in which every twenty seconds or so "mental handicap" transmitters send out distracting noises to keep people "from taking unfair advantage of their brains." In the story, sounds from gunshots to crashing cars keep the characters from completing a thought or a conversation and possibly gaining an advantage over another person. In real life, e-mail, tweets, and telephones keep us from finishing a sentence or getting our work done.

José Saramago, the late Portuguese novelist and playwright who won the Nobel Prize for literature in 1998, offers a different analogy in his novel *Blindness*. Saramago's story opens in a world just like the one you know today. People are living their lives, building a career, driving to work, starting and raising families, attending to errands and meetings. Then a person sitting in traffic in a car instantly becomes blind. The blind man is rushed to a doctor, who in turn soon becomes blind.

Quickly and efficiently, blindness spreads through society as an airborne virus. The government mobilizes and begins to quarantine anyone showing signs of blindness. As people are corralled and moved into hospital-like prisons, one group isn't fazed by this epidemic: those who were blind before the epidemic started. They take over camps full of the newly blind, who are inhibited by the new way they are forced to navigate the world. The already blind become the leaders of the new sightless society.

The blind, once at a terrific disadvantage in a sighted world, now have a terrific advantage. For them, blindness is nothing new. They know how to get around, how to cope, how to navigate a world that no one can see anymore.

In my mind's eye, I see my ricochet work style and that of young workers who text on their phones, type on a computer, share videos and images, listen to music, and talk all at once as much·like the blind in Saramago's work. Our new way of working, once a disability, now has the potential to be clearly valuable. Already, you often see job descriptions with prerequisites such as "must have the ability to multitask," which essentially translates to "Can you do ten things at once?" A quick search of the word "multitask" on the online job board Monster.com generates thousands of results asking for people who can manage X, Y, and Z at the same time.

To me, it seems feasible that the members of a generation growing up doing their homework while simultaneously engaging in a number of other activities will come into the workforce and engage with their office duties in the same way. It is not unlike earlier generations coming into the workforce and displacing the pen with newfangled typewriters and then displacing the typewriter with the personal computer. But is this just wishful thinking? Is our jumping from task to task truly efficient—an unappreciated ability in the modern world—or merely engaging enough to lead us to think we are accomplishing plenty when in fact we are mostly spinning our wheels?

The result matters a lot today, when we are wirelessly connected to anywhere in the world. Every year, the gizmos we carry in our pockets can do more and more things, encouraging us to take advantage of them not just when we have a break but while we're walking down the street or driving. The temptation is to jump at every beep and buzz of the cell phone and every ding of the computer inbox, to answer every communication right away, and, for many of my generation, to search for the answer to every question that pops randomly into our heads.

But we already know there are some huge risks with this impulsive behavior, particularly when we combine just about any cognitive task with driving, which requires alertness and quick reaction times. Although I tend to engage in multiple activities when I work, I would never do that while operating a vehicle. As my colleague Matt Richtel at the *Times* wrote in 2009, the Virginia Tech Transportation Institute put video cameras in the cabs of long-haul truckers and watched over eighteen months as the drivers talked and texted their way from location to location. The findings: Texters' risk of a collision was twenty-three times greater than that of those who were simply driving. Another study of college students in a driving simulator found that the young people were eight times more likely to crash while they were typing on their phones.

Is this a problem of learning and practice? Does it matter if you are young or old, tech-savvy or geek-averse? Is it possible that we could build gray and white matter so that we can efficiently handle these various tasks safely at one time? Or is our wiring such that we truly cannot master several cognitive tasks in tandem? If so, will we need to schedule them the way we do with the gym or television shows, setting aside, say, Twitter time outside of our work or our driving in order to devote our attention to them?

Yet because we can't text and drive, does that mean we can't chat with friends online or text while doing homework or other tasks? Does that also mean that we can't consume truly multimedia storytelling, watching videos, interacting with graphics or images, leaving comments for friends, and consuming the information in a thorough way? I know this is the way I work, and quite successfully.

To find out if I was the exception to the rule, I went on my own quest, consulting leading neuroscientists and cognitive psychologists on the human potential for multitasking. I hoped that by corralling the work of these scientific experts, I could share their knowledge as it applies to the changing media landscape and whether we will have to change the way we tell and consume stories. So I asked them, Sure, we can walk and chew gum at the same time, but can we productively type, talk, and read all at the same time? And does it make us more effective or creative?

The Cocktail-Party Problem

The thorny problem of multitasking has been a workplace challenge for some time, I learned, dating back more than half a century, when commercial air traffic started to increase rapidly. In the early 1950s, air traffic controllers faced a serious problem. Airplane traffic was soaring and the controllers were handling a growing number of planes in the sky. But many control towers, sometimes with several people managing multiple planes, operated on a single loudspeaker. Information about individual planes came in all at once, a cacophony of crucial information that was hard to decipher. Pilots would begin their descent to airports and announce their flight patterns by radio to the control tower. But the messages from individual pilots merged, and controllers had to decipher this blended jumble of monotone voices while trying to guide planes in for safe landings. It was more and more difficult for controllers to follow a single plane amid the alphabet soup of call letters coming in.

"North Tower, this is Boeing 737 Alpha with a Mercer Departure at Alpha Niner Delta. Altitude 400 feet, moving at 383 knots." This type of jumble came from multiple planes, sometimes at the same time. There was a lot of information about one flight for a controller to absorb, and, worse, the potential for disaster was enormous.

In the 1950s when Colin Cherry, a well-known British cognitive psychologist, heard about this problem, he began to wonder how people generally distinguished between multiple voices, such as the individual voices at a party. A field of research developed around what came to be known as the cocktail-party problem.

It's a great question: How do people at a noisy cocktail party hear their own names called out by a friend or easily converse with one person while ignoring the noisy discussion of surrounding guests? The question that Cherry and other researchers explored was: If you can hear your name being called out and engage in discussion at a noisy cocktail party, why can't an air traffic controller decipher between two audio inputs simultaneously?

To study the cocktail-party problem, Cherry decided to test a number of different audio problems. For the first set of tests, he recorded one person reading two different texts and played both of them at the same time to individuals to see if they could differentiate one from the other. The subjects were asked to listen to one of the messages and separate the two different topics they heard. The outcome showed, Cherry wrote in the 1950s, that although "the results were a Babel, nevertheless the messages may be separated." The people were able to focus with one ear and allow the other ear to push the competing content aside—perhaps much the way a parent carries on a conversation with one ear while constantly listening with the other ear to children's play (or fights) in another room.

Cherry performed numerous variations on this test, using different languages, phrases, and accents to determine when pairs of voices were distinguishable and when they weren't. In another series of tests, he put headphones on people in the hope of directing one message to the right ear and another to the left ear. And as the test progressed, he gradually changed different values and inputs while encouraging the participants to try not to listen to one ear and focus on the other, just like at a cocktail party.

At first he tried a barrage of wacky ideas, such as playing audio into the left ear in German spoken by an Englishman. In this test, subjects were asked to decipher what they heard. Cherry later experimented with accents, switched between male and female voices, and even reversed playback of prerecorded audio. Certain attributes went completely unnoticed. Some were noticed quickly by most participants.

He theorized that certain factors help us differentiate between multiple sounds, including the direction the voices come from and the visibility of people's lips. Other distinctions included things as simple as a male or female voice, subject matter, accents, and a difference in pitch.

Cherry didn't figure out the inner workings of the brain and how it is capable of paying attention at a cocktail party while flushing out the unimportant pieces of chatter. Instead, he figured out how we filter this information. Cherry discovered that a variety of factors help us distinguish and filter a tremendous amount of auditory information. Seeing someone's lips move is a perfect example. Accents, pitch, and the direction of the voice all play other crucial roles in deciding what our brain will process. Although Cherry found that it was impossible for most participants to consume two conversations simultaneously, he found that the brain is partially capable of paying attention to other audio inputs even if it doesn't process and remember all the information.

Some years later, as research progressed in this area, key experiments found that people are more capable of understanding multiple sound inputs when the inputs are drastically simplified. For example, if people hear the word "bread" in one ear and then hear the somewhat expected word like "knife," which would make up "bread knife," in the opposite ear, they can comprehend through both ears. But if they hear "bread" in one ear and something completely off topic from the word "bread," such as "carburetor," in the other, they'll be much less likely to comprehend or retain the duplicate conversations. These later experiments, which were performed by the psychologist Donald Broadbent, showed that "messages containing little information can be dealt with simultaneously, while those with high information content may not." Or as Broadbent said in one of his research papers that was cited by MIT computer scientists, "the statement 'one cannot do two tasks at once' depends on what is meant by 'task.'"

Research around the cocktail party problem was initially aimed at helping computers understand sounds, which still hasn't been perfected, not to solve the mystery of multitasking.

But sixty years later, researchers are still trying to completely understand the cocktail party problem and what is actually happening in our brains as we hear multiple sounds. Even in 2005, a paper in the MIT Press journal *Neural Computation* noted that "it seems fair to say that a complete understanding of the cocktail party phenomenon is still missing and the story is far from complete; the enigma about the marvelous auditory perception capability of human beings remains a mystery."

What doesn't remain a mystery, and what research about the cocktail-party problem tells us, is that our brains somehow can discern multiple inputs at once. Our ability to multitask is not a binary question of yes or no. It depends drastically on the task at hand. The fact that we can't drive a car and send a text message safely at the same time doesn't mean that we can't engage in multiple conversations in chat windows online or even consume a new kind of book that includes audio, video, and commentary. As these studies show, if the content relates, its parts can be consumed at the same time and might even be able to tell a more engaging story.

Blink. Don't Blink.

The cocktail party problem was first researched nearly sixty years ago. Brain research has since been catapulted into the mainstream, and there have been many thousands of new studies and findings on the inner workings of the brain. To understand the multitasking debate, especially when it comes to storytelling, I found I needed a better understanding of how the brain works. I was told over and over by many neuroscientists that first and foremost researchers still don't know a whole lot about what goes on between a person's ears. As Richard Haier, who performed the Tetris studies, noted: "One [thing about the brain], it's really exciting to do brain research, and, two, we don't know anything about the brain." One neuroscientist pointed out that we still don't know or understand how my brain can tell my hand to pick up a glass of water and bring it to my lips.

That said, we are starting to understand small bits and pieces of the brain and how this applies to the future of storytelling. The following studies help paint a better picture of how our brains work in some of these scenarios.

In the early 1990s, Jane Raymond, a professor of psychology at Bangor University in Wales, wanted to understand how our eyes and brains work together and how well they process information. Working with other researchers, she discovered that human brains can move only so fast before they factually miss something. At certain speeds, the brain just doesn't process information sent by the eyes.

Raymond and her colleagues named the phenomenon the attentional blink. These blinks aren't information that is missed by the eyes as they send messages to the brain. Rather, the brain itself actually appears to blink.

Raymond used a testing process called RSVP, for "rapid serial visual presentation," which shows shapes or letters in rapid succession, fast enough that the images change up to ten times in a second. She found that at certain rapid speeds, the brain misses the next image. It doesn't even register the event. It's as if the brain were actually blinking.

Researchers in neuroscience labs around the world have

studied attentional blink over the last two decades to try to understand the significance of brains missing bits of information and seeing certain content only when it is delivered at a limited tempo. One key conclusion is that some tasks truly limit our brains' ability to do two things at once—though they may be able to do two things in very rapid succession, so quickly that you could hardly tell that the actions didn't happen simultaneously.

Paul Dux, a cognitive psychologist now based at the University of Queensland in Australia, in particular, wanted to know if we could train our brains to move faster, just as video games can improve our response times and awareness.

Dux describes the brain as a "pretty advanced processing system between our ears," capable of doing amazing things— even tasks that a computer may never be able to perform. At the same time, he notes, we have severe impairments. "If you're driving a car and trying to talk on your cell phone at the same time, you simply can't do that successfully," he says. "We also find it very difficult to tend to two visual tasks, or interact with more than a couple of objects at a time."

Most of the time, he has concluded, "you just can't perform multiple tasks, even if they're very, very simple."

But, he wondered, maybe we've just never been asked to perform these kinds of multiple tasks at once. Basing his hypothesis on past research, he asked, "If we practiced multitasking, could we become more capable? Could we improve our abilities?"

Working with another neuroscientist, René Marois of Vanderbilt University, Dux asked participants to try to perform two very simple tasks simultaneously. For example, they showed participants one of two colored disks on a computer screen.

The subjects then were asked to press the right hand's index finger when they saw one color and press the middle finger when they saw another one. While the participants were paying attention to the color disks on the screen, they were also asked to listen to differently pitched sounds and notify the scientists when they heard a high or low pitch.

Dux and Marois found that although people absolutely could not perform both functions at once, they could, with repeated training, improve their ability to multitask and increase their speed and accuracy in processing information. In fact, participants actually improved their rapid switching abilities almost tenfold with continued training and practice over a few weeks. Dux and his colleagues were able to do this by essentially training the prefrontal cortex region of the brain, which is responsible for processing these multiple tasks, to work faster.

Of course, there are true human limitations to how far we can adapt, and for some people it's always going to be easier than it is for others. Some of us have a very small attentional blink, whereas others show a large attentional blink. In a controlled lab setting, people with small attentional blinks are good at identifying symbols and letters quickly in studies using rapid switching. People with large attentional blinks have trouble identifying the second element.

Dux says that there are other key differences between these groups: "Subjects with smaller attentional blinks are much better at inhibiting distracting information. They can essentially suppress information that isn't relevant to the task at hand. If a random image or color appears in the screen, they will simply ignore it." But people with a large attentional blink become easily distracted when trying to perform a focused task. "Thus, the blink affects not just the processing of the person but also how the person copes with distractions," Dux said.

With these distractions, people with smaller attentional blinks just "ignore and suppress them. It's not that they don't process them at all. They actively inhibit them." That's a subtle difference to point out and an important one in the understanding of how people process information. Dux theorizes that "it might be that the information initially gets in, but they're very good at keeping that information out" that will interfere with other tasks.

That skill is sometimes known as the executive function: the way the brain organizes, plans, schedules, and handles distractions as well as multiple tasks. When a person's executive function is humming along in good working condition, that person can stay focused on a job and push away impulses and distractions that might interfere. Not surprisingly, the skill has become a hot topic in schools, where exercises and lessons that improve executive function may help students learn more quickly and make them better at subjects like math.

This cognitive skill, which also helps us surf the Internet while watching TV, seems to emanate, according to the scientist John Medina, from the prefrontal cortex, the home to Brodmann's area 10, our ever-important multitasking switch. Medina is a developmental molecular biologist who has focused a great deal of his research on the genes associated with the development of the human brain. He is also the author of a very colorful and fun book about the brain called *Brain Rules*. In a recent interview with me, Medina explained that the capability of the brain for consuming simultaneous pieces of information lies with Brodmann's area 10. Animated about the potential and the limitations of Brodmann's area 10, Medina cautions that multitasking isn't necessarily the most productive way for people to work. He explained that each time we switch tasks, it costs us about 700 milliseconds of brain power, not a lot when you do it once or twice, but over an eight-hour workday, it adds up.

That 700-millisecond number goes back to a 2001 research paper written by Joshua Rubenstein of the Federal Aviation Administration and David Meyer of the University of Michigan. Rubenstein and Meyer were studying the effects of multitasking on pilots who have to pay attention to multiple inputs at once, including augmented information on the screen.

I asked Medina and Meyer in separate interviews if this number could be increased, or does every human work exactly the same and can Brodmann's area 10 shift rapidly shift between tasks only at 700-millisecond intervals? They both said that although modern research hasn't proved this yet, it is probably true that the next generation's task switching works a little more quickly. They both also said that maybe a brain like mine, which has grown up with computers and video games, could rapidly shift faster, maybe even as fast as at 350 milliseconds between tasks. But even if that is the case, Meyer cautioned, we do know you are eventually going to hit a ceiling limit. We can flip back and forth only so quickly.

Medina also noted that although it's fine for people to engage in this kind of "rapid shifting" in a social setting, he believes it can have negative effects on the brain in serious professional settings, slowing us down or wasting valuable time when we are constantly switching tasks. In other words, multitask responsibly.

Multitasking Mirage

Medina, like Dux and Raymond, insists that our brains are
processing just one job at a time—quickly maybe, but still just
one thing at a time. "We can speed up the tempo of switching,
but our brains will never be able to do these tasks simultaneously," he says.

But, I wondered, we still seem to multitask. So how is it,
I asked Medina, that as a kid, I grew up listening to my headphones while doing my homework or reading a book? If I sit in
a quiet room trying to write, I easily become distracted. If instead I've got some music with words and lyrics playing in the
background, I can sit and work happily for hours. Now, I can't
concentrate unless I'm doing those two different tasks at once.

Medina explained that I've become accustomed to working
in this way, what he called "state-specific learning." The music
is essentially like white noise within my brain, distracting my
distractions to help me concentrate. In other words, my brain
has adapted to incorporate both of these things; though I'm
really focusing on the task at hand, the music is mostly background noise.

What I did as I grew up is also quite similar to the way
people have adapted over generations as new, more distracting technologies have come along. With each new technology,
consumers have to figure out how to add it to their lives. They
have to decide when they want to read or listen or watch. For
most people, these new experiences don't destroy a previous
experience. They only fragment our current media consumption.

Clifford Nass, a Stanford University professor, developed what he calls "partial displacement theory" to explain that as new media such as television and the Internet develop, they don't immediately replace older ones; we simply "displace" the new medium and meld it into our current habits. For a long time, many of you probably kept a cassette player in your car and used a CD player at home. Later, you probably had a CD player in the car and an iPod in your pocket. People didn't stop listening to the radio when television came along; instead, they found a new time and place to consume the older medium. And as one kind of medium is displaced, it begins to overlap with others.

Just think about how many media are in our lives: magazines, newspapers, movies, TV shows, thousands of websites, friends' chats or text messages—the list could go on and on. But there's only a certain amount of time in the day to consume all of this. We have to work, we have to eat, we have to sleep.

Mechanizing the printing press during the Industrial Revolution produced far more printed material than the world had ever seen, forcing us to make choices about what we had time to read. (It was probably no coincidence that the Sears, Roebuck catalog was frequently found in outhouses in the early 1900s, where it provided reading material and the pages offered other uses in a pinch.) Radio, which became widely available during the 1920s, did not put an end to people reading books, newspapers, and magazines. Instead, it changed how much time we devoted to print experiences.

You've no doubt seen the images of a family sitting around the living room: Dad, Mom, three kids, all happily looking up at a giant box—staring at the radio. People would sit raptly listening to a radio program for an hour with full undistracted concentration. At first radio station options were limited. Then more stations and types of programming appeared, and we gradually started to listen to more radio. Soon, as more shows and options appeared, the "one hour of radio" in the evening turned into two hours, then three, and soon people stopped staring up at that radio and instead looked down again, reading newspapers and books while listening—or, in a fashion, multitasking.

When the television arrived in a big way after World War II, it didn't replace the radio, which until then sat comfortably in the corner of the living room, although plenty of people speculated that it would. The television did, however, change where and when we used the radio. Families now, for the most part, watched TV in the living room for a couple of hours a night and listened to the radio in the car, a technology that first became available in the late 1920s.

Although the television didn't become widely watched for some years to come, it still signaled a new form of media for us to cram into our daily diet. In turn, this led to more time that would be taken up by news, information, and entertainment. What better way to handle all these forms of storytelling than to start consuming content in places people didn't think they could engage with? Coupled with the time limits in a single day, they began to move through all at once. People listened to the radio while reading a book or watched TV with a computer on their laps and—voilà! They were media multitasking.

Rather than decide between a newspaper and radio, consumers chose to do both at the same time. Or rather than decide between surfing multiple websites on my laptop, watching a TV show, texting with a friend, and playing a video game, I'll just do them all simultaneously. The coming generations will figure out even more consumption combinations and collectively will most likely become even more adept at juggling different types of media.

Our brains may be switching back and forth in milliseconds, but from a public perception, we seem to be growing accustomed to the changes—or at least comfortable with them. And although many scientists can't agree on the negative-versus-positive aspect of this task switching, scientists, psychologists, and communications theorists all seem to agree on one thing: There's no turning back the clock. Whether we want to call it multitasking or context switching—and whether it's good for society or bad—is somewhat irrelevant at this point. We're all engaging in multiple activities at once. That said, there is one solution that can help curb this rapid, unrelated multitasking, and that involves better, more immersive storytelling.

As research around the cocktail party problem (see pages 203–207) illustrates, our ability to register and process the task at hand can be more productive and useful if the tasks our brains are processing are related to one another. If content creators, teachers, or parents want to keep the people of the next generation engaged, they will need to create storytelling that takes advantage of their multitasking minds in a way that can relate to the information they are consuming. They need to learn how to talk to a generation that is "easily distracted" and whose "mind wanders too much." For example, rather than only giving me the opportunity to send text messages and tweets while watching a documentary, why not create an experience in which my computer can call up additional information such as Wikipedia pages or commentary from other viewers, thereby creating a fluid multiscreen experience?

Generation Multitask

Perhaps no group has embraced multitasking more than the young, those now in high school, in college, or in their early twenties. In a 2006 look at "The Multitasking Generation," *Time* magazine science editor Claudia Wallis profiled how junior high, high school, and college students jumped from medium to medium, instant messaging and working on homework while iTunes played on the computer or even flowed into an earbud stuffed into a single ear.

Researchers were shocked and surprised by the youngsters' overwhelming need to indulge in several tasks at once—to the exclusion of family dinners or even pleasant conversation. They saw it as the biggest change in family dynamics over the last couple of decades. We also see it with adults. In the middle of a meeting or lunch they pull out their BlackBerry or iPhone to check an e-mail and tell you, "I'm listening."

The young people appeared to be impressively good at juggling multiple media at once. According to a study by the Kaiser Family Foundation, the time young people spent with media was holding steady with previous surveys at six and a half hours a day. But by IM-ing or listening to music while they watched TV or worked on the computer, the young people were fitting eight and half hours of media exposure into that period. (The study referenced in the article was released in 2005; these numbers have continued to grow rapidly since then!)

In the *Time* article, Pier, a fourteen-year-old boy, explains how he does his homework: "I usually finish my homework at school, but if not, I pop a book open on my lap in my room, and while the computer is loading, I'll do a problem or write a sentence. Then, while mail is loading, I do more. I get it done a little bit at a time."

Some of us know that feeling well. Your computer has twenty tabs open on several browsers. You've checked e-mail in the middle of an instant message exchange. Then you've popped over to an article to try to read a few more lines before heading back to something else. You've gotten pretty good at it, right?

But are you and all those young people really better at multitasking? For all the neuroscience studies proving that we can perform multiple tasks better with practice, there are some communications studies that say it's not practical to switch between tasks. For example, a recent paper published by Eyal Ophira and Clifford Nass for the National Academy of Sciences suggests that maybe you're fooling yourself.

Nass and Ophira are both researchers at Stanford University in the Communication between Humans and Interactive Media Lab. Nass, currently the lab's director, has spent his career looking at the effects, both positive and negative, that computers and media have in our lives. Byron Reeves and Nass's book *The Media Equation: How People Treat Computers, Television, and the News Media Like Real People* looked at the advent of the television age on our culture.

When Ness and Ophira began, in 2009, to study whether multitasking made people better on cognitive tests and in memory skills, their assumption was that those who moved easily from job to job would perform better than those who burrowed into a single task, just as practice had improved some of the dexterity and responsiveness of video-game players. The team of researchers assumed that people who engaged with multiple media were actually better at keeping out distracting information.

The tests used in the research involved showing the participants a series of red and blue rectangles on a screen. The participants were asked to ignore the blue rectangles as they flashed around a screen and pay attention to only the red ones. The subjects who said they were low multitaskers had no problem ignoring the blue rectangles. Those who said they were high multitaskers were distracted by the blue rectangles. These tests were performed using letters and speed, too, but each time the low multitaskers performed better than the high multitaskers.

The results were surprising: "Heavy media multitaskers performed worse on a test of task-switching ability, likely due to a reduced ability to filter out interference from the irrelevant task set." In other words, heavy media multitaskers were much worse at concentrating than were light media multitaskers. Nass explained that the heavy media users were easily distracted and actually slower.

Still, all the researchers I interviewed agreed that we can't place a stake in the ground when it comes to multitasking until there is more research. Even Nass, whom I spoke with several times about his and others' research, said that it will take years before we know the reality and limitations of our brains in a multitasking society.

A study released in 2010, just one year after Nass's multitasking research, by two University of Utah researchers showed that a small segment of society is truly capable of multitasking. The research involved two hundred college students and their ability to talk on a cell phone while using a driving simulator. Nearly all of them failed miserably—not surprisingly. But a very small number—2.5 percent—had a "super-tasking" ability to drive and perform other tasks without any decline in results. These extraordinary multitaskers even repeated their strange skill on a second test. Unfortunately, there were few clues to help the researchers figure out which drivers would have superior skills—and many people assumed they were one of the rare few.

Although young people—say, those under twenty-five—may seem to be more attuned to this kind of switching than their parents, multitasking isn't just a generational characteristic. L. Mark Carrier and Nancy Cheever from the Department of Psychology and the Department of Communications at California State University, recently surveyed 1,319 people who were split into three different segments based on age: baby boomers (born between 1946 and 1964), Gen Xers (born between 1965 and 1978), and Net Geners (born after 1978). The survey asked questions related to the experiences they engage in simultaneously, such as listening to music while playing video games and texting or e-mailing while watching TV.

The researchers found that some tasks just don't mix, regardless of age. That is, very few people said they play video games and chat on instant messenger screens at the same time. And as you might expect, very few people read books for pleasure while texting or e-mailing. But the study showed a very high level of multitasking across all generations. The researchers pointed out that some multitasking is mindlessly easy, no matter how old you are; for example, all generations could listen to music or eat along with other tasks.

Carrier originally hypothesized that the majority of media multitasking took place among the younger generations. He also believed that this younger group would be much better at performing any two tasks at once. Instead, the researchers discovered that everyone engages in multiple variations of media simultaneously, though baby boomers found more tasks difficult to perform simultaneously.

Carrier also discovered that many of the difficulties combining tasks were similar across age groups. For example, multitasking while reading for pleasure was the least likely to happen simultaneously (although 46 percent of the Net Geners reportedly tried to do it often anyway). That's not surprising in light of the depth of thought that reading requires. When you read, "you recruit more of your senses, you recruit more of your higher-level reasoning processes, your imagination gets more involved. If you do it right, it's a mentally intensive task. It requires paying attention to the material and accessing your long-term memory," Carrier said. A lot of the information in the text requires you to infer and draw that conclusion in your imagination. All of that makes it very hard to read while sending texts or answering e-mail.

But one side effect may be that more difficult tasks, the ones that really require your brain to kick into third or fourth gear, may be less appealing. Carrier says that research shows that "traditional reading, print reading is not as engaging" to certain younger groups anymore. That's not to say all reading or all groups of young people. But once students have had the chance to experience multimedia approaches, they may find the results much more interesting. Younger kids, who are exposed to that kind of stimulation more often, now think and work with different kinds of visual and auditory stimulation. "They didn't grow up thinking print reading was the end all, be all of the highest level of scholarly pursuit," he said.

Carrier's points about reading are at the heart of the debate over multitasking. Kids come home from school and open their laptops (purportedly to do homework) but may also watch movies, chat with friends, or update their status on a social network. Then, when they sit down with a book, their brains say, "Hey, wait a minute, I'm not used to just sitting here with words. Where are the images, where's the conversation, where are the pop-up windows?"

Reading is an extremely engaging task, and if done correctly, it can engage the imagination and numerous areas within the brain. Reading also forces the brain to think deeply, wiring our brains for deep introspection and thought. It's also an imperative part of growth for the brain and ingenuity. But this doesn't mean that all forms of reading and learning need to happen this way: There's a balance of other forms of media that we can swirl into the learning apparatus of the brain.

Innovations in electronic books may well change the way we look at reading in the future. A history book about, say, the Civil War could include a video game instead of just having words and maps. After reading about the Battle of Gettysburg, for example, you might go into the battle as a soldier or a general and experience this turning point of the war "firsthand."

Or a biography of Albert Einstein could include an interactive avatar-like program of him. You could ask him questions about his life or about the theory of relativity. You could engage in an interactive conversation with an actor or read his papers

video
[Book concept]

together. To me, that sounds like a pretty compelling form of learning.

This is the type of stimulation and learning the next generation may demand. In the media survey for the Kaiser Family Foundation, a seventeen-year-old girl explained, "I get bored if it's not all going at once, because everything has gaps—waiting for a website to come up, commercials on TV, etc."

As we will see in chapter 8, the experience will drive the success of future stories. People who make their living telling stories will feel more and more pressure to create experiences that offer multiple layers of content, additional social feedback from a community with shared interests, threaded topics, and true interaction. If they don't, they may capture only their audience's partial attention.

From a scientific and research-based perspective, Carrier believes that as you add more simultaneous media to the way we teach and tell stories, you'll "recruit more of your senses, you'll recruit more of your higher level kind of reasoning processes. Your imagination gets more involved, and you get higher motivation levels."

From a personal perspective, especially as I think about what I learned while researching this book, I believe Cheever is right. My own exploration included conducting interviews, watching videos, listening to lectures, and reading research papers and books. I created my own form of interactive learning. Future students and researchers will do more by expecting their sources to be archived and searchable and available in multiple formats. And if the story is told in a fashion that the multitasking generation is used to, they will give the subjects more than one ear, or at least their partial attention.

City Multitasker/Country Multitasker

All the studies discussed above show how rapidly our brains are capable of adapting to and melding with new environments. Some of these changes are iterative, happening as new technologies enter our lives, and some are new and explosive, but our arguably underutilized brains just morph and readjust to the new experiences.

If Johannes Gutenberg had invented the Internet five hundred years ago instead of inventing the printing press, our brains wouldn't have exploded and turned into sloppy green goo. We would have figured out how to use the new technology and harness it to share information and tell stories, just as we are doing today.

Do we create the technologies to cater to our brains' thirst for stimulation, or are our brains just doing what they need to do to keep up? Most of the scientists I've interviewed agree that the brain's thirst for stimulation drives the technological advances of each new innovation. We want to know more, and we want to see it, smell it, feel it, and hear it, engaging all of our senses in the experience. Young people growing up are getting a taste of this in their own learning and their own exploration, and they will want more, both for themselves and for their own children.

Reading and imagination remain important. But how can we expect a child who goes online three or four hours a day, pecking and clicking, defining his or her media path, delving into an immersed and fully interactive storytelling experience, to sit still and read a book or watch a movie if the experience is not stimulating her brain properly? Some will say, of course, that these children are spoiled (or stupid) and that they have lost the ability to concentrate. Or some may assume they have ADD and shouldn't spend as much time online since it only compounds the problem.

One initial response is to limit the amount of time spent playing video games, the hours online, or the number of text messages. It's a mistake to think that this ricochet behavior is a "problem" that needs to be "solved." The problem isn't the multitasking generation but the media they are consuming. What if we look at it from the other point of view? Maybe these older types of content—books, movies, television, and news-papers—aren't adapting appropriately to the technologies and expectations of the young and old, of today's adapted and more demanding brains.

Video games and the Internet are not a detriment to our brains and society. Learning how to manage fourteen buttons at once or navigate content-rich websites is a benefit, not a hin-drance to further learning. As I've spelled out, research shows that video-game players have superior eye-hand coordination, an increased capacity for visual attention, and a vastly superior set of spatial visualization skills.

This doesn't mean that all books and TV shows need to become storytelling carnivals full of color, noise, and moving pictures with embedded crawlers along the bottom of the page. There should be a balance, and the result should be relevant to the content and the viewer who is consuming it.

John Medina emphatically explains in his book that no two brains are the same. He cites Michael Jordan, who is consid-ered the best basketball player of all time. Jordan's brain is built and attuned to basketball more than that of any other human being on the planet. But as Medina points out, when Jordan decided to start playing baseball full-time, he was literally the worst player in the league.

The same thing applies to the way we consume media. "The bottom line," as Richard Haier said to me, is that "if you think of the brain like a thermostat, some people have their thermostat for stimulation set very high and others have it very low. So you may love rock music, but you may hate going to concerts because you find them way overstimulating—too many people, too loud—even though you appreciate the music. Or think about people who enjoy a quiet weekend in the country. They find it relaxing and stimulating. And other people, city people, can't wait to get out of the country and back to the city because they're not stimulated enough."

The beauty of the next ten years, as more and more content types begin to move permanently onto screens of all shapes and sizes, will be the ability to pick the experience that's right for you—to engage in the type of stimulation that fits your hyperpersonalized preferences.

If you want to consume a more prosaic type of storytelling, that should be your option. If for you, as for me, that wouldn't be stimulating enough to your brain, an immersive supplementary experience should be available to you. And if the content creators don't tell the story in this new immersive way, you may well be able to create a substitute yourself.

It won't have to be all or nothing. People who live in the city still like driving in the country on weekends—even if they drive a little faster than the people who live there year-round.

8

what the future will look like
a prescription for change

The future is already here—it is just
unevenly distributed.
—William Gibson

What the Future Will Look Like: Dinner on the Moon

On the run from the police in the science-fiction movie *Mi-nority Report*, the character played by Tom Cruise decides to take cover in a Gap clothing store. There he is greeted not by a happy, breathing Gap employee but by the digital avatar of a helpful clerk. In a flash, the see-through saleswoman recognizes him via an eyeball scanner and instantly remembers his recent purchases.

"Hello, Mr. Yakamato!" she greets him. "Welcome back to the Gap!"

"How'd those assorted tank tops work out for you?" the digital sales associate asks.

The scene is just sixteen seconds long, but it has achieved near-cult status among advertising executives, designers, and technology nerds. In one respect, this moment of the movie is both comical and realistic. Through this quick exchange you get an exciting and maybe frightening glimpse of the future. Implicit in that brief meeting of the eyes and the scanner are all the possibilities of an entirely new way of shopping. But even more alluring is the potential for radical new daily experiences.

Ultimately, that's where all this technological upheaval is taking us: to a world rich in new and different experiences. Already, the Web and digital devices have changed how and where you read, watch, and listen and what you read, watch, and listen to. They have changed the communities you interact with. They have rearranged your brain cells and the way you think about everything from maps and locations to friends and relationships. They have shifted your approach to the world from a third-person perspective to a first-person one—and a hyperpersonal one at that. Most of this sea change has bubbled up from users as they brought the new technologies into their lives and adapted as the technologies changed them.

Now the companies must figure out how they will adapt and sell products in this shifting environment. As was fictionalized in *Minority Report*, it will be up to the Gaps, the Starbucks, the automakers, newspapers, book publishers, and moviemakers to decide which technologies to adopt and how to use them to their fullest advantage in moving their products, shows, and content. Ultimately, some companies will win, and those winners will be the ones that create the best and most meaningful experiences for their customers.

A lot of visionaries and futurists worked on the *Minority Report* concepts. Steven Spielberg, the director of *Minority Report*, asked his team of designers to envision what the year 2054 might look like. Spielberg tapped into the creative talents of famous writers such as Douglas Coupland and Stewart Brand, and also worked with interface designers from the Massachusetts Institute of Technology, including John Underkoffler, the movie's science adviser.

A creative team involved in the retail experience of the movie said "customers did not actually have to try on the clothes in the store but could do so in a virtual way." A three-dimensional representation of your body would be stored in your mobile phone or wristwatch. The information would then be transmitted to a life-sized "virtual mirror." Dale Herigstad, a designer on the movie's retail concept said, "You could then see specific sizes and styles of clothes on your virtual self in the mirror" and even put yourself in different environments, such as a park or office space, so that you would have an idea what that red dress or blue suit would look like at a dimly lit party. Then you could send images of the setting to a friend and ask if "my butt looks big in these jeans."

Some people might find this all a little creepy and invasive. A computer system would know who you are, when you bought your last T-shirt, the true size of your belly or butt and even if they had changed since your last shopping spree, and the socks and underwear you prefer. Already, there have been some initial efforts. For a time, Levi's made "perfect fit" jeans based on a person's measurements, but it stopped in 2004 when it closed its last domestic manufacturing plants. Lands' End offered a digital look at how clothing would fit if you entered your own sizes. But those attempts were crude compared with the possibilities of digital technology today, which can account for your unique curves and angles, coupled with settings that can even tell you what your date might be wearing and how he or she will look. Imagine if you could carry your exact information and preferred styles on your phone rather than trying on thirty different styles of jeans to find the most comfortable and flattering fit.

Other ideas that were presented, but not used, in *Minority Report* are also in nascent stages, such as wallpaper that is actually a flexible screen. The team once mocked up a restaurant setting with those digital walls. If you want to eat in Venice in real time, you tell the restaurateur, and voilà, your booth is flush with Italian gondolas floating by as you sit along the canals. Or if you prefer a chicken sandwich for dinner on the moon, no problem; that exists, too—and you don't even have to change into a space suit. Or maybe you're in New York and a family member is in LA. The two of you could eat dinner together through the virtual wallpaper. Of course, you couldn't pass the ketchup, but you could enjoy each other's company and feel as if you were in the same room together.

Other concepts that didn't make it into the final edit of the movie included an "oasis" that would allow some respite for people experiencing information overload. Designers imagined an option to pay to enter a totally controlled space to relax and completely shut out the information chaos outside. There you could reboot your mind in a controlled setting that matched your interests aurally or visually. Beach lovers could experience the quiet calm of the Caribbean for an hour, and those who preferred the mountains might do the same at the top of Mount Everest. Concepts like these aren't meant to replace a trip to the beach but instead are meant to give you a calm break from the unstoppable flow of information in an ever-more-connected world.

Not only are we starting to see experiences like this that bring us into other worlds, we're also seeing them happen at a much quicker pace than Spielberg imagined in his 2002 movie. We're seeing a society in which our lives are augmented by our ever-smaller and more powerful mobile devices and our online preferences accompany us wherever we go—even into the real world. The next challenge is to convert these technological abilities into profitable businesses that serve the consumnivore's growing appetite. Of course, that's much more easily imagined than accomplished. But the good news is that just as the porn industry showed us in the beginning of this book, it's not one size fits all; a number of products are likely to fill the bill.

What the Future Will Look Like: Does Screen Size Matter?

Whenever people find out what I do for a living, they ask the same questions: First, how long will paper be around? Quickly followed by, Which device will replace paper for reading books and newspapers? Will it be the Kindle? The Nook? The iPad? Or something yet to be announced or even maybe envisioned? Will it be flexible? Will it be as small as a deck of cards or as big as a broadsheet newspaper?

I hear the question everywhere I go—at conferences, at dinner parties, even around the office. People inevitably want to know which specific device will solve all the problems that will come with phasing out paper-based media. For a long time, I have to admit, I thought one gizmo would hold all the answers as well: "one device that rules."

One of my first tasks at the *Times*'s research and development lab was trying to answer this question and assess what the next generation of the newspaper might look like: what might become the iPod of electronic readers. Part of my job was to keep track of upcoming new reading devices and screen technologies and to understand where digital readers and screens would be in the next two years to two decades. It was not an easy task, but it was definitely a fun one.

The best part of the job was ordering new gadgets. We tracked just about anything with a button and power source: motion-tracking remote controls for your TV that allow you to navigate your living room like a video-game controller. Completely flexible screens that respond to multiple fingers at the same time. Virtual keyboards that wirelessly link to your phone or computer and visually project keys onto any surface you want to type on, from a desk tabletop to a sidewalk. Microprojectors that are the size of the end of your little finger and can project an incredibly bright display up to thirty inches wide, which is larger than most standard monitors. Then there were the e-readers, devices such as the Sony Reader, Amazon's Kindle, the Apple iPad, and book applications for mobile phones, along with a wide range of funky-looking European and Japanese reading devices that were never intended for the U.S. market.

My desk was a constant reminder of the speed with which these devices were being developed and marketed. I sat amid a clutter of opened boxes, packing slips, and bubble wrap. On three long tables behind me sat almost every e-reader device made in the last ten years. We referred to the pile of buttons, screens, and power cords as the "gadget buffet."

Playing and experimenting with these wacky and innovative inventions helped us alert people in the company to the perpetually changing world of devices. They also helped us understand the direction of the marketplace and how each product would affect news content. For instance, if people were to start reading news on their televisions, we needed to be ready for that transition to a bigger screen and figure out how to organize and display *New York Times* stories accordingly.

Though my colleagues and I often saw eye to eye on what worked and what didn't, we couldn't agree on the shape and size of that just-right reading device. The more I listened to questions during Q&A sessions or heard from people at cocktail parties or conferences, the clearer it became that a variety of devices might be needed, some with buttons, some with touch screens, some flexible, some rigid, some big, some small. Each person seemed to have a different preference. I came to the conclusion that just as we have TVs that range from pocket-sized to bigger than many New York City apartments, we probably will need a plethora of different kinds of readers too.

The assumption is that bigger screens will be better, but that may not be the case. Cheryl Bracken, a professor at Cleveland State University, has spent the last decade studying the way we process media content, focusing on whether screen size and the quality of a screen are really relevant in viewing content. Why, she wondered, would someone watch a movie on a 2.5-inch iPod screen when she could sit on the couch and watch it on a 42-inch TV? But that was just an assumption based on her personal experience. Bracken wanted to understand if the next generation, the digital natives, felt the same way.

She and her researchers recruited ninety-eight undergraduate students to test their viewing experiences on different kinds of screens to find out if one experience was more enjoyable than the other and whether the viewer's understanding of the narrative is lost when a story produced for a large screen is viewed on a small screen.

The students were shown two different clips from a movie on both an iPod with a 2.5-inch screen and a TV with a 32-inch screen. Each clip was approximately ten minutes long. One clip consisted of longish, slowly developing scenes; the other clip was much faster, with rapid edits and a high-speed car chase.

In theory, at least, the participants would find watching the movie on a larger screen more engaging and immersive than using the smaller version. But in fact, the end results were significantly different. Students who watched the film on the iPod found the experience almost twice as immersive and engaging as those who viewed the clip on a larger television.

Why? Bracken says the study found that the headphones used with an iPod effectively closed out the rest of the world, helping viewers focus more intently. Further, the subjects holding an iPod felt a greater sense of control over the storytelling and watching experience because they literally held the experience in their hands. Holding a device in your hands allows you to move the device to fit your viewing preferences; a large TV mounted on a wall requires you to move to accommodate it. In other words, screen size, sound, and comfort weren't the defining factors in the experience for these digital experts. Surprisingly, control over the process and experience, the ability to tune out distractions, and an immersive experience turned out to be hugely important.

That doesn't mean that tiny devices are the answer for everyone. One person who doesn't have a desire to watch movies on small cell phone screens is David Lynch, a director who has been nominated for twelve Oscars and directed some big-name movies, including *Blue Velvet, Twin Peaks*, and *Mulholland Drive*. Not only does Lynch not want to watch movies this way, he thinks anyone who chooses this experience won't get the full effect he intended when he directed the movie. During a recent interview on television, he indignantly scorned people who watch movies on their phones when he said that "now if you're playing a movie on your telephone, you will never, in a trillion years, experience the film. You'll think you have experienced it, but you'll be cheated." After a brief pause, he yelled into the microphone, "It's such a sadness that you think you've seen a film on your fucking telephone! Get real!"

OK, we can assume that Lynch won't be watching the next Super Bowl on his iPod. But that's the beauty of these digital experiences. I'm perfectly happy watching *Mulholland Drive* on my iPhone. Lynch might prefer a movie theater. You might be comfortable somewhere in the middle, sitting on your couch at home or watching on your laptop. Digital affords options and preferences, not generalizations.

But there is a ceiling to these small-screen devices. One limitation of small gadgets comes from the optical accuracy of our vision, what scientists call human visual perception. When screens or fonts or details are very small, our eyes strain to see them clearly, often without success. Eventually the size affects our attention span. That's why we get headaches from reading really small print or looking at something with too much detail for a long period.

So if people can enjoy a movie on a two- or three-inch iPod screen, how small is too small? Can you happily watch the latest episode of *Entourage* on a screen the size of a postage stamp? What about your thumbnail?

Researchers at the University of Portsmouth in England had the same question when it comes to students and learning. At first they wanted to understand if mobile phones could be used in schools for teaching. And if so, was there a cutoff point where the diminished size started to affect the experience? The researchers chose a group of young children in school and tested them on what they learned from the supersmall screen.

The students were shown different videos on mobile phones with three different screen sizes and then tested to see how much information they retained. The largest screen was a little less than four inches wide, the medium was about the size of an iPod, and the small was a little more than one and a half inches wide.

One of the videos the students were shown illustrated how to fold a piece of origami. Afterward, the students were asked to try to perform the same task from memory. The students viewing the instructions on the medium and large screens retained significant amounts of information, and the screen size didn't affect their learning or memory of the video or their enjoyment of the exercise. With the smallest screen, however, the students had just as much fun watching the video as they did on the other two screens, but their ability to recall information from the screen was much lower.

Nipan Maniar, who led this study in the United Kingdom, said that research consistently shows that students who watch educational videos on medium or large screens retain significantly higher amounts of information. He sees the mobile phone, with its midsize screen, becoming an integral part of the classroom over the next ten years; teachers will be able to hand out coursework wirelessly, communicate one to one, and even allow the students to learn in a highly personalized manner that might incorporate video, reading, multimedia, and games.

Sound familiar? It would be a classroom for Me!

Every person's mind is built completely differently, totally unique from that of everyone else. Asking twenty students to read the same textbook at the same time is like expecting that group of students to be able to run a mile at the exact same speed or to have an equal ability to paint a still life. Our brains are simply not built that way.

Using screens and digital teaching will allow kids to engage at their own pace in a collaborative fashion that paper just can't provide.

What the Future Will Look Like: 1, 2, 10

Although smart phones are now the rage, a large proportion of my work over the last several years has revolved around these mobile devices. And with good reason: By the end of 2009 there were nearly 4.6 billion active mobile phones in the world. With the entire global population at 6.6 billion, that means the penetration rate for mobile could be as high as 70 percent. (Some people own two phones.) And we take these little gadgets with us everywhere, slipping them into and out of our purses or pockets several times a day. As they've evolved, so has our dependence on them.

Several technologists, myself included, believe that the mobile phone probably will outpace desktop computing in the next five years as the central entrance point to the Web. But the mobile phone doesn't signal the demise of the desktop computer or the large television screen sitting in your living room. Instead, these Web-enabled devices will start to talk to one another and interact in ways that might seem like science fiction today.

At New York University I teach a course on this topic called "1, 2, 10." These simple numbers represent the distance a screen is from your eyes. Cell phones and e-books are approximately one foot away when you hold them in your hands. Computer screens are about two feet away. The average television in the living room is, you guessed it, ten feet away. The idea of the course is to explore how content can automatically follow you from screen to screen and place to place, and with this experience the content can automatically change and adapt between these different devices and a person's locations.

The 1, 2, 10 concept presents incredible challenges. Designing interfaces for a television screen, where you're usually sitting ten to fifteen feet away from the image, is a completely different challenge from designing an interface for a mobile phone that is about the size of a chocolate bar. As I teach my students in the class, on top of these vast differences in size, it's imperative for consumers to switch seamlessly between these experiences without even realizing they moved the same experience to a different screen.

computer - tv - phone.

Imagine if you're reading an article about a new food recipe on your computer at work. When you get home from the office, your television should know that you've read the article and automatically show you video clips of the recipe on this new screen. At the same time, as your phone is in the same room with you, with the flick of a button the television can send the recipe to your mobile phone so that you can pick up ingredients at the grocery store the next day. If you want to take this one step further, you can imagine your fridge notifying your phone which ingredients you already have for the recipe.

I believe that technology that responds to your precise location at the moment will be in the next wave of products that we start to soon see enter the electronics marketplace, allowing for more customization and personalization of information, entertainment, and advertising. For instance, if I am reading the newspaper at four p.m. on a Friday in the Park Slope section of Brooklyn, the content I see should reflect the time of day (near dinner), the place (what's nearby), and more. The news feed I'm reading should know what I've already read that day and what I haven't. If I don't like sports, I shouldn't see articles about sports. It should factor in what my friends have read and what's being discussed on my social networks. Most important, these systems should do this without my having to instruct them or tell them anything.

In the same vein, whatever you're watching or working on could stay with you, moving from computer to phone to television or actively appearing in a different context on all three if you prefer. Consider the frustration of my friend Michael whom I worked with in the *Times* research labs. Michael showed up to work one Monday morning, and when

I asked how his weekend was, he explained that it was a little frustrating. He told me that he was watching the final innings of a baseball game when a friend invited him to join him at a bar a few blocks away to watch the rest of the game and share a few drinks. Michael wanted to see the friend but didn't want to lose the thread of the game. "I really wanted the content to follow me, for my phone to know that I was leaving my house and to know that I was watching the game on my TV," he said. "My phone should know all of this and send me updates as I walked to the bar. When I arrived at the bar, my phone should be aware that I am back in front of a TV and stop updating me with the scores."

It's not an unreasonable idea or the utopian fantasy of a technophile. In fact, Michael and I decided to build a rudimentary version of a similar experience. But instead of a baseball game, we used *New York Times* news articles as our muse. A mainstream version of this technology doesn't exist yet, so we had to do a little tinkering and hacking.

To start, we took a cell phone, placed an RFID chip inside, and then attached an RFID reader to our computers. An RFID (radio-frequency identification) chip is a tiny electronic chip that can store little pieces of information that can be transferred wirelessly to a RFID reader device that interprets the identity of a chip. Many businesses, mine included, give cards with RFID chips to employees so they can enter their office buildings without using a key. RFID chips are also in some credit cards so that you can wave your card in front of an ATM machine instead of swiping it through a scanner. Using these chips and our mobile phones, Michael and I were able to let a computer know we were there just by placing our phones on the desk.

It was a simple hack to keep track of a person's presence and location: Place your phone at the desk, and the computer knows you are there. Pick up your phone and walk away, and the computer knows that you left. Using this detection, Michael and I wrote some code that kept track of the articles we were reading on NYTimes.com and could automatically pass the articles back and forth between the phone and the computer without our having to do anything. So if you're reading an opinion article by Nick Kristof and you're halfway down the page, we know you are not finished with the article, and when you walk away from your desk, the rest of the story will automatically appear on your phone. We conceptualized scenarios that would take this further. Imagine if you got into your car and your phone automatically started playing the audio of the article or if you came home and a 3-D avatar started reading the rest of the piece to you on your television.

Right now, a lot of this is wishful thinking. First, many of these devices still aren't connected to the Web. The TV is on a cable network, the mobile phone is on a cellular network, and the computer is connected to a separate Internet provider. But when all these experiences move to the same network, the Internet, they can easily start talking to one another. Even now we are starting to see a new wave of cars that are connected to the Web and can notify you via e-mail when it's time for an oil change.

This three-screen concept has been on its way for years. I can check my e-mail from my laptop and my phone. If I delete an e-mail on one of those devices, it will be deleted on all of them. I can listen to music on my TV, laptop, music player, or phone. But right now, I have to load the music separately to make that happen. What Michael wanted was for his phone to actually talk to his television and vice versa. Just as millions of people are paying $25 a month for Internet services on their smart phones, Michael's wishful thinking is another example of the kinds of experiences people would willingly pay for if they found the results useful and valuable in their daily lives.

Coffee shops + Newspaper now + school

What the Future Will Look Like: People Pay for Experiences, Not Content

Daily, we see examples of great experiences that people are clearly willing to pay for—important, eye-opening investigative stories in the form of nonfiction books or newspaper articles; absorbing movies that bring people in droves to theaters; mind-blowing music concerts; moving novels; and of course porn.

Often, you don't even need special technology or an unusual innovation: There's that coffee I buy at my favorite Brooklyn café, where I pay for consistency and convenience. In other instances, it's adding something to an already existing product: Some *Times* readers pay concertlike prices to attend the *New York Times*–created lecture series built around some of the paper's best-known writers that bring in sold-out crowds. I pay for the *New Yorker* magazine, which consistently offers enthralling prose no matter whether I experience it in print form or digitally. For kids, this can come in the form of a traditional television experience melded with new media. Take the *iCarly* show, the hottest thing on television for kids and tweens, which uses a filming technique developed by MTV in the late 1980s to produce a fast-paced, engrossing show. Quick cuts, multiple angles, and sometimes a first-person point of view in which the screen is meant to look as if the viewer is holding the camera help keep young viewers involved, as does social networking. Just like stars of blue movies who chatted with viewers and shared little details of their lives, the teen characters on *iCarly* talk to fans online through social networks and the show's website, continuing the story and conversation with their audience long after the thirty-minute episode has ended.

Given how easy it seems to be to identify something that rises above the ordinary, why is it so infuriatingly difficult to figure out the right kinds of great experiences that incorporate and make full use of new technologies? If great content can be made meaningful, why does future revenue still seem so murky for so much of the media world?

Consider the unfolding battle between the many technologies and approaches to book publishing. It seems pretty clear that some time in the future, paper will fall by the wayside, becoming more expensive to produce and distribute than digital screens, and a good many of us, if not most of us, will read books on some kind of gadget. But with so many publishing companies experimenting with digital books, the best experience—or even a really good one—is far from clear.

Although we don't know what will work, a world of Me Economics and the shrinking cost of divergent hardware options probably will mean there will be a choice of reading devices to fit your preferences. Take the approach so far from online booksellers. Amazon.com originally took the low road, offering a simple black-and-white reader and a big inventory of electronic books on the assumption that simplicity and price would be the main drivers. With its $9.99 price tag on most books, it actually loses money on almost every sale, according to *New Yorker* media writer Ken Auletta. Amazon believes that a low price will build market share and consumer loyalty. Already, Auletta said, Kindle readers buy far more books than they did when they were purchasing print alternatives.

One reason for increased digital book sales is the price, but the other factor provides more evidence that people pay for experiences, not just content. The purchasing experience on the Kindle is seamless, simple, and instantaneous. Let's say you hear about a new book from a friend. You can navigate to the Kindle bookstore through the device, and your new book is in your hands minutes later.

But Amazon doesn't sell just books in its online store. It also sells magazines and newspapers, yet the number of subscribers is surprisingly small. The exact numbers of magazine and newspaper subscribers for the Kindle are not made public, but as *Time* magazine reporter Josh Quittner wrote in May 2009, "The *Wall Street Journal* is the second best-read newspaper and has sold a mere 5,000 subscriptions to date." An internal memo that was leaked on the Web from the *New York Times*, which is the number one selling newspaper on Kindle, said that the *Times* had 10,000-plus subscribers. Although I was unable to find the exact numbers, a source at Amazon told me that the highest newspaper or magazine subscription for all three Kindle devices combined is in the mid-tens of thousands. So why are book sales so high and sales of other publications so low? Because the newspaper and magazine experience is terrible; they are selling only the content. Amazon allows publishers to distribute their content only into a booklike layout.

There are no images, no typography, no consistency with the brands, just text on a page.

Other players in the early e-reader space included the consumer technology giant Sony, which tried to jump into the e-reader business by promoting comfort and ease of use but fell short on both. The Sony Reader never managed to make the inroads that Amazon has because the entire experience was flawed. The first generations of the product required a USB cable to move books to the device, and the company didn't announce digital newspapers until December 2009. I owned first- and second-generation Sony Readers, and getting books out of the proprietary Sony bookstore was utterly painful.

Now Amazon, Sony, and other players may fall behind Apple's iPad and a wave of e-readers built using Google's mobile Android operating system, which both offer book reading as one of dozens of applications.

Apple, mimicking its iPod experience, has aimed for the high end, assuming that a color screen, a very fast response time, and the "cool" factor will help it grab market share in books the way it has in music. The iPad, at least initially, sold for twice the price of a Kindle, and electronic books in Apple's iBookstore sell mostly for $14.99, a price that pleases many publishers but sets up a duel with Amazon.

When Steve Jobs, Apple's chief executive officer, presented the iPad to an audience of six hundred geeks in January 2010, he talked about consistency, simplicity, and a uniform interface. Jobs walked his audience through the simple experience of buying and then reading a book and while doing so talked about those three points in detail. He explained: "We've created the new iBookstore, fully integrated with the iBooks apps, to allow you to discover and purchase and download e-Books." As Jobs sat in a black chair on stage, he ushered the audience through the iBooks application and navigated through the store features. He explained that "if you've used iTunes or the App store, you're already familiar with this [interface]." Jobs then purchased a book that downloaded onto the device instantly. The entire four-minute demo probably doesn't sound that exciting, but for Jobs and Apple, that's the point. They don't want people to have to think about anything other than the decision to make a purchase. The rest should be a seamless and simple experience.

The search engine giant Google, which has been busily scanning millions of books for the last few years while also navigating a copyright lawsuit with authors, joined the scrum. It is selling works from a Google electronic store called the E-book Marketplace that will be readable on any device, including e-readers and mobile phones, and can also be sold through bookstores. A Google executive I spoke with for a *Times* article said that the company hopes to use its prowess in searching to create a flawless experience for customers.

The shopping experience for the e-reader device is only one of the challenges booksellers need to grapple with. There's also the way the story is told. Consumers yearn for more interactivity and the better kinds of storytelling that are afforded by color screens, multitouch interactivity, and social interaction with friends. In some instances, content creators will have to experiment and engage readers, or viewers, in new ways.

In the end, all these companies are going to be on the same footing. Google, Apple, Sony, Amazon, Barnes & Noble, small digital booksellers, and even some publishers will all offer books directly to consumers and will all be selling the same content. Both booksellers and book creators will have to figure out how to offer a better experience for consumers to entice them to come to their stores—either with a different kind of adventure around the purchase of a new book or with the additional and supplemental type of storytelling offered on these newfangled devices.

It's impossible to predict what will be the draw for an individual customer. Some may base a purchasing decision on price alone. Others will be drawn by the ease of the purchasing experience, the level of interactivity within the story, or the extended life of a narrative. Still others may base their decisions on immediacy alone. But one thing's for sure: The content is only one tiny piece of the puzzle.

What the Future Will Look Like: A Storytelling World, with Participation

As the big guys wrestle with which products will produce the best and most significant experience, people like me will keep experimenting with what already has been created, demanding immediacy, personalization, networking, and easy access. I'm an early adopter of technology and excitedly embrace and try any early technologies I can get my hands on. For some, it may seem like I live in the future. Before too long, though, you will be there with me. Or as the science fiction writer William Gibson once said: "The future is already here—it is just unevenly distributed."

I realize that although these technologies are creating some amazing changes in the way we live and work, they've also up-ended entire industries and generated a great deal of fear and anxiety. Among the points I hope you will take away from this journey into what's ahead is that such fears are a normal part of adapting to radical changes in the way we live. It's understandable to feel unsettled and disoriented. But again and again in our history, we have adapted and moved forward, and in doing so we've learned to tell better stories than previous technologies allowed. We survived and flourished as trains replaced wagons and cars replaced horses; as radio and then television brought information directly into our homes and then fought for ownership of the living room; as comic books, video games, and iPods provided new forms of entertainment. As a society and as an economy, we will survive and then flourish amid this new influx of fast-moving information as well.

In addition, be reassured that storytelling will remain a central part of our lives. We may tweet in 140 characters, but more and more of those missives already are delivered with links to photos, videos, and stories. In other words, they've become essentially personal headlines with more detailed information attached. And even when we all transition from paper to pixels, we will still read book-length content and consumer news articles written by people who are paid a salary to share a 1,000-, 5,000-, or 70,000-word story.

Long-form content will not disappear even if we consume it in forms different from paper, even if it comes with embedded videos or with sensors and augmentation as part of the narrative. And people will still pay for all these forms, with significant and meaningful content as a crucial part of that experience.

Working in the newspaper industry, I'm completely aware of the constant level of anxiety from both my colleagues and my readers over the fate of news. The uneasiness is apparent and real; newspapers have been going out of business at alarming rates, leaving the question in the balance: What is the future of news, and does it exist?

I believe that there will still be a number of news outlets and a news business in the future—though they will look drastically different from the way they look today. Some of these organizations might be increasingly specific or personal, serving the needs of a relatively small number of readers rather than a mass market, sort of like in the olden days. Before newspapers and journalists as we know them, individuals were paid to be professional correspondents for rich merchants and powerful clergy. In the sixteenth century, those correspondents were sent to other cities to gather information and send letters back to their benefactors detailing shipping news and prices. The first newspapers, then, were private intelligence for individuals.

As the first rough newspapers started to take shape, people were still part of the conversation. It is believed by some historians in England that several of the first newspapers encouraged readers to write their thoughts on the pages before they passed a paper along to another reader. It wasn't until the eighteenth century that publishers began to sell news to the broad public.

For news to remain relevant to consumers in the future, many newspapers and news organizations are going to have to adapt and change. The theories behind the role of news shifted dramatically in the 1920s as two writers and thinkers, Walter Lippmann and John Dewey, entered a growing public debate about the role of the newspaper in society. Lippmann argued that the public was unable to govern itself properly. Instead, he believed, experts, or journalists and those in government, were needed to tell the people what they had to know. It was their job to explain science and politics to the masses. He argued that workers had plenty to worry about trying to pay their bills and put food on the table. And most important, they didn't have the time or even the knowledge to ask informed questions about government or society. Lippmann essentially argued that the role of a journalist was to tell people what they need to know and what to think about it.

Dewey, in contrast, argued that the person who wears the shoe knows where it hurts. He believed democracy worked only if people understood the problems their country faced— and that newspapers and journalism were a perfect vehicle for that conversation. Even if the masses were limited in what they understood, he thought the job of educated people, communicators, and journalists was to use their best tools to engage people in the news as participants. Essentially he said, Let's allow the people to work with journalists and tell them what to report.

For the most part, Lippmann's theories won, mostly because the people who owned the newspapers and the printing presses concluded that the role of journalists was to tell people what they needed to know, not to hold a conversation. But today, the pendulum is swinging back in the other direction. With the advent of social technologies such as blogs, comment opportunities, Twitter, Facebook, YouTube, and other simple sharing tools, the masses have gained a collective voice on an unprecedented scale. The public now has an equal voice with the printing press and no longer wants to sit idle as mainstream media dictate the day's news. The result may be a change in the way news is reported and told in the twenty-first century, a process that may become more conversational and more personal for those who want to participate in the experience. That's an evolution that would be perfectly consistent with the history of newspapers.

Other types of news will come in the form of bytes and bits. As more information becomes available from our digital devices and sensors, we will see reporters emerge from sensors and algorithms. Open government initiatives and the creation of websites such as data.gov are forming hubs for the government to share information and data to be used in stories and information gathering. We are entering an era of news reporting that will blur the line between algorithmic news gathering and storytelling with human curation and explanation. Stir in social aspects and shared voices and you've got a perfect mix of Lippmann, Dewey, computing, and the general public.

What the Future Will Look Like: Undercutting Yourself

For newspapers and other media businesses, the changes have been wrenching, and some news outlets have lost ground to technology companies that aggregate news, such as Google and Yahoo!, which are more nimble in posting news as it happens. Responding quickly to the changes can be at odds with responding thoughtfully, and some companies end up paralyzed by the challenge. But with tastes and technology evolving rapidly, the ones that hesitate may truly be lost and the ones that move aggressively may win the game.

Look at Apple, the early computer company that has moved into music, music players, cell phones, and new electronic readers. In 2007 Steve Jobs, Apple's chief executive officer, had to decide whether the company should introduce a new product that could drastically hurt the sales of a successful current product.

For almost thirty years, Apple's bread and butter was selling personal computers, related software, and peripherals. But in 2001 Apple introduced the iPod, a small music player that eventually would change the entire shape of the music industry. By 2006, the iPod accounted for the majority of its core business. In late 2006, Apple reported that it had sold an astounding 21 million iPods in the last quarter of the year. The iPod business and iTunes combined brought in $4 billion of the company's $7.1 billion in revenue for the quarter. In comparison, its Mac computer sales accounted for $2.4 billion in revenue. You would think that Apple would do everything it could to keep those iPod revenues. But the company had other plans.

Apple recognized that music players were eventually going to simply be additional pieces of software built right into a phone or another device. So in 2007, Jobs stood on the stage at the Macworld developer conference in San Francisco and made two announcements: First, the company was changing its name from Apple Computer to simply Apple—a clear recognition of the company's metamorphosis. Second, Apple was introducing a new product line: the iPhone.

Jobs explained to the crowd of wide-eyed geeks that this new sleek shiny gadget wasn't just a phone. Sure, it would make phone calls (though not especially well, thanks to the

AT&T network). It was also designed for e-mail and surfing the Web and had a mapping application and a calendar, and by the way, it had a free iPod stuffed inside.

This was a risky move. Customers who purchased an iPhone surely wouldn't need an iPod, too, and the phone would definitely cannibalize the sales of the company's core business. But Jobs knew that if he didn't move ahead from the iPod, another company would.

The move paid off. In the first quarter of 2010, Apple announced that its revenue had jumped to $13.4 billion, nearly twice the total from late 2006. Apple grew dramatically after the iPhone launch. In early 2010 its market capitalization was a staggering $222 billion, surpassing its biggest rival, Microsoft, as the world's biggest technology company. Although the company sold 10.9 million iPods, half the number it sold three years earlier, it also sold 8.75 million iPhones.

Jobs knew that if he didn't undercut himself, someone else would. There was an incredible amount of risk involved in subverting his core business with a new product, but this is a philosophy Jobs understands from the early days of computing when Apple lost the computer wars with Microsoft, the dominant force in the computing world. Innovation has clearly played a huge role in the company's rise since Jobs returned in 1996. But that innovation has been coupled with a willingness to make a popular product obsolete, creating one of the most profitable and successful technology companies in the world.

The same challenge applies to other business industries too. Some newspapers, magazines, book publishers, and music houses are trying to hold on to their cash cow—their paper or plastic products. In the interim digital-only companies are springing up out of nowhere to compete without the same infrastructure, expenses, or business traditions.

What the Future Will Look Like: TMI?

Vinton Cerf, considered by many to be the "father of the Internet" and now the chief Internet evangelist at Google, has a message about your socks.

During a presentation at Google several years ago, Cerf explained that some day, everything will be connected to the Internet. That includes a person's socks, so if one falls behind the washing machine, it will be able to notify him, or the other sock, of its new location.

In Cerf's vision—"the Internet of Things"—sensors eventually will be everywhere, embedded in our T-shirts and the medicine we take, and will be able to deliver real-time information and analysis to our persons.

In a blog post I wrote about this topic for the *Times,* I explained that we're already seeing the beginnings of this: "Doctors are using tiny cameras, about the size of a pill, to look at the digestive tract and send back information and pictures. Farming equipment can collect data from remote satellites and sensors in the ground, anticipate weather, and adapt the fertilizer to be used. And billboards in Asia can change displays based on the preferences of passers-by."

Understandably, the Internet of Things, as it is called, scares some people. Embedding the Internet into everything could make us reliant on technology that could crash at any moment.

But even more, it means that even greater masses of information will be created, much of it increasingly personal and unique. These technologies raise new and difficult questions about privacy and appropriate use of what we know. Some of the folks who live much further into the future than me highlight this challenge.

For example, if you bumped into Steve Mann at any point in the past few decades, you would definitely remember him: He looks like a cross between a computer and a human. Mann is considered one of the first digital cyborgs and has been experimenting with wearable computing for the last thirty years. He invented a patented system he calls EyeTap, which he says should be used "for electronic newsgathering, documentary video, photojournalism and personal safety," where the wearer becomes part of an "intelligence network."

When I first met Mann at a conference a few years ago, he was wearing a huge pair of goggles that looked more like a Halloween costume than a computer and seemed to shield his vision completely. The goggles had a series of wires that were connected to his scalp and also connected to a computer attached to his waist, which monitored information about him and his surroundings and made the information visible on a computer display embedded in the goggles over his eyes. Mann called this setup "mediated reality."

For conference attendees, Mann plugged the system into an external projector so we could see what he sees. At the time he was eating lunch, and the screen filled up with a picture of some peas and other greens, all surrounded by a series of charts and numbers. Mann's heart rate and other vital information were displayed. His wearable computer was also recording all the sounds and images in the room and uploading them to the Web.

At first I was enthralled with the idea. How amazing would it be to augment your reality with a device like this? You would never forget where you left your car keys or how to say "hello" in another language.

Then I met Gordon Bell, a seventy-five-year-old researcher at the Microsoft research labs in Seattle, who several years ago created a device called the SenseCam, which sits around his neck like a large necklace and records every aspect of his life, taking up to a thousand photos a day. He also records audio of every interaction, just like Mann. Everything he sees is wirelessly sent back to his computer and available for retrieval at a later date.

Mann, Bell, and other cyborgs who capture their lives continuously not only push the limits of what someone wants to know about you but also generate sincere worries about whether there are life events better left out of the record. True, it's possible that having so much information inscribed somewhere else may free up our heads for more creative and productive thinking, as one expert told the writer Clive Thompson when he profiled Bell in *Fast Company* in 2006. But Frank Nack, a German computer scientist, noted that he was a big fan of forgetting, which is crucial to forgiveness, moving on from setbacks, and even appreciating nostalgia.

"It's how we make sense of life, how we interpret things," Nack told Thompson in the *Fast Company* article. "Everybody is building a life story; we all need to forget certain stages. I don't want to be reminded of everything I said."

My excitement about Mann and Bell's retrieval systems hasn't completely disappeared. There's still a part of me that would love to walk around with an augmented vision of reality, but I recognize that there needs to be a balance in the information we collect. There needs to be a way to opt out of the constant retrieval of images, audio, and information. When I met Mann, my image was automatically recorded for later use. The only way to avoid being under his surveillance was to run away. What do we do when the Internet or computers refuse to forget? How will we cope in the future with political candidates who left stupid photos on their high school Facebook pages or sent a drunken college tweet that any thirty-year-old would regret?

I recognize how important this would be from my own colorful past. Although I grew up on the Web, thankfully, there were no social networks and digital cameras when I was in my early teens. Instant messenger exchanges, happily, weren't saved the way Gmail chats are today. That's a good thing for me, because when I wasn't on the Web, I was out with my friends, getting into trouble.

Thankfully, those exploits weren't on Google when my career was gearing up, although they will be once this book is published. When I was thirteen, I was arrested for stealing a pack of cigarettes, but since I was a minor, it doesn't show up on my record. When I was fourteen, I got in trouble with the police for graffiti. That's nowhere to be found, either. At fifteen, I was suspended from school for fighting. (I lost, of course.) That's not on Facebook or Twitter.

If Twitter, Facebook, MySpace, YouTube, or other social networks existed when I was twelve, you can bet I would have bragged about my experiences to my online friends—as I did in real life back then. And those details would still be around on the Web for anyone to find. If those records existed online when I joined the workforce, it's possible I would never have been hired by the *New York Times*.

All this is a cautionary tale for the future. The Web and technology need to leave a place for people to make mistakes. They need to allow for youth to make mistakes. While holding people accountable for genuine wrongdoing, they also need to have some room for anonymity and for forgetting so that young people—and even some older ones—have room to grow and change.

That perspective is shared by Christopher Poole, founder of the message board 4chan, where people can anonymously post responses to just about anything, often using the entire range of four-letter words and pornographic images as well. Though he acknowledges that some posters say vile and disgusting things, he believes the people who come to his site have a right to do so anonymously without sharing any personal information. They have a right to make mistakes. Poole doesn't keep any personal information about his users, and after a certain period, all the posts on 4chan disappear like products on a conveyor belt.

When I talked to Poole for a profile interview, he told me about a recent technology conference he attended at which someone else defended anonymity, saying, "Part of the magic of youth is that people are able to forgive and forget." On the Web, Poole said, there's a chance that kids will never have that opportunity to make mistakes, to forgive and forget, unless some parts of it remain anonymous and short-term. "As kids, we say stupid things, and because there's not a record of it, nobody is going to give you a hard time at thirty years old about something you said or did when you were eight years old. Online, you have all these social networks that are moving to a state of persistent identity, and in turn, we're sacrificing the ability to be youthful," he said. "In ten years, everything you say and do will be visible online, and I think it's really unfortunate."

Currently, there is no statute of limitations for stupidity or immaturity. Today's young people will have a harder time dismissing their misbehavior as President Bush did, saying, "When I was young and reckless, I was young and reckless." But our future will be much harsher without some understanding that what happens in the online world shouldn't always stay there forever.

You can be sure that Mann, Bell, and today's cyborgs offer a glimpse of the future for a distant generation. Our mobile phones and digital cameras already record millions of photos each day. Just as it's important that websites such as 4chan exist, even though most of society won't agree with their content, it's going to be equally important that certain aspects of the future allow us to forget pieces of the past.

What the Future Will Look Like: More Personalized, More Possibilities

If we don't all self-destruct, what will be next for us on the technological front?

Well, everything, actually.

The "Me!" concept isn't just about your news being personalized. It's about everything being personalized, from the bytes from your computer and mobile phone to the full meals of your home and life. Imagine that you could get a personalized flexible digital newspaper and each time you turn it on, it delivers only the news that's relevant to you, based on what your friends have read, where you live, and other individual interests. That's not too far off.

Now imagine that the same thing applies to objects. Maybe you're having a large dinner party and need two extra Asian-themed plates and cups to match the set you already have. You could just print them out. Or perhaps you want a collar that can tell you where your cat is and send a message to your phone if he gets lost.

This sort of object and hardware revolution is now under way, mostly in the garages and workshops of hobbyists, just as computers were the dreams of tinkerers in the 1970s and early 1980s. A few years ago, I began to tinker with building my own electronics and started meeting people online who were also interested in understanding how a transistor or microchip works. I started to meet with other electronics hobbyists once a week to share projects and help one another solve problems. As word got out, the meetings grew. Eventually, we rented a workspace and became an organization called NYC Resistor.

The entire purpose of NYC Resistor is to make things. We are hardware hackers—no, not the kind of hackers that break into bank accounts and shut off power grids but the kind that turn one kind of hardware into another. You can think of it as a fight club for nerds, but we try not to punch each other.

Just like the home-brew computer clubs of a generation ago, there are other nerdlike fight clubs cropping up all over the world in which people build all kinds of crazy contraptions. At NYC Resistor, the group worked together to create a robot called BarBot that can pour and mix alcoholic drinks. Another member of the group takes old iPods and turns them into drum kits and other miniature music-making machines. Another group member, Diana Eng, makes clothing with built-in electronics that make her outfits sing with glowing LED lights and futuristic materials. Her outfits blur the line between fashion and functional clothing.

I recently built a "smart" lamp, a four-inch opaque cube that sits on my desk and can glow in different colors, depending on news alerts that I've set up. If Barack Obama proposes a new bill and it's picked up in the news, the lamp glows blue; if there's news in my neighborhood in Brooklyn, the lamp will glow orange. It's not a very practical application, but it's a product I wanted, so I decided to build it. In the future you will be able to build your own personalized products too. Two other cofounders of NYC Resistor might be able to help. Zach Hoeken and Bre Pettis, whom I would characterize as nerds times ten, started a company called MakerBot where they build and sell "open-source 3D printing robots." Imagine a printer sitting on your desk at home that can actually "print" objects in plastic.

MakerBot is a kit that can be purchased and assembled for around $500. Once it's together, you can download schematics from the Internet of anything from a mousetrap to a cup and actually print them out in plastic. By contrast, a low-end 3-D printer today costs about $20,000.

There are other companies being built out of this personalized hardware concept too. Bug Labs, a small computer hardware company in New York City, sells a device called a BUG that comes with a variety of "modules" that plug into one another. The main base of the BUG is a little computer about the size of a deck of cards, and the various modules are half that size, a couple of inches square. Say you want a device to monitor your children at the playground while they are with the baby-sitter. One way to do this would be to make a gadget that takes a picture every ten minutes, checks its location, and then e-mails you the photo and a map. Bug Labs's idea is to let you make it by taking a BUG computer and adding a camera module, a GPS module, and then an SMS module to give your new contraption access to the Internet. Using your computer, you program it to take the steps you want. When you're done, you have your own personal long-distance kid monitor.

Although most of this is some years away and the home hardware hackers are largely a group of nerds like me, some day we may all have 3-D printers and other plug-and-play technologies that allow us to create objects personalized for each of us. It's an exciting proposition.

For some people it might also be a scary one. We don't know exactly what these things may look like, whose jobs they may replace, or what the impact of instant manufacturing may have. Just as the emergence of a connected digital world has come with bumps and bruises as well as revelations and surprises, the additional advances that bring us new capabilities will come with problems and hiccups as well as unexpected developments that rattle our world.

epilogue
why they're not coming back

Dear CEO, Publisher, Producer, Editor, Author, Journalist, Advertising Director, Filmmaker...

They're not coming back.

Traditional consumers aren't coming back. Print advertising isn't coming back. Media, brands, and the established narratives aren't coming back. And almost everyone will eventually make this transition.

I'm not going to wake up one day and say, "Hey, the Web isn't for me; I'm going to start buying CDs, print books, and newspapers again." I'm among the new era of consumers and contributors, and we're looking for new forms of content and storytelling. Where it doesn't exist, we're going to find it elsewhere, make it ourselves, or, in some instances, just take it.

I'm not alone in this thinking. I know part of you hopes that these changes are going to stop or at least plateau. But they're not. This isn't just a temporary bump in the road. This is society changing before your very eyes. Just as the printing press helped cement and form imagined communities that became nations, the Internet is doing the same thing, changing our concept of location, trust, space, time, and connections.

Sure, the irrational economy has affected the speed with which this has all happened, it's forced us to push fast forward on the demise of the DVD player, newspapers, cable television, and most things analog. But I can assure you, they're not coming back.

Before you panic any further, be assured that first and foremost we're all driving off this cliff together. The entire business of storytelling—music, movies, television, newspapers, books, public relations, advertising, teaching—every business will be affected. We are all going through the same involuntary mutation. Some of us have already left solid ground, and others are heading toward the impending ledge. But one thing is for sure: We're all going over that cliff. What happens at the bottom of the ravine is what we get to decide, and for some of the luckier ones the lessons of others will help us prepare.

You see, when you take it down to its core, we're all just storytellers. Whether you're writing a book or a news article, selling an outfit or a car, writing a blog post about your weekend or a press release about a new product, you're telling a story. Whether it's 140 characters long, the length of this book, a video, interactive, 3-D, or an in-person narrative, it's a story.

In the past stories cost money and were told by people with

access to a printing press or television studio, but now everyone has the ability to spread and share information equally. With inexpensive tools at our disposal, with our mobile phones, digital cameras, and laptops, we all have an equal voice. A short mobile video clip of a riot in Chicago, uploaded to YouTube by a passerby, sits alongside a video from a multimillion-dollar television network like CNN. A tweet sent by a student in Iran can reach the same number of people as a message sent out by the *New York Times*.

In addition, the anchoring communities all of us are creating— our social networks—help ensure that each message is filtered and shared with equal importance and fairness and reaches each of us in an individual way.

The consumers who aren't coming back are scurrying like ants in every direction possible, and you're probably wondering where they're going. They're searching. Searching for new forms of storytelling that we haven't offered yet. The bottom of the ravine, the new medium, affords a new narrative—just like in the early days of television, when the producers didn't know what to do with cameras and motion and started filming radio shows. The business of storytelling is doing the same thing with the Internet. We're taking our existing content and simply aggregating it to the Web; we're filming radio shows.

As unsettling as it may sound, we need to accept that we are not simply selling content. We're not selling the words on the page or the images on the screen; instead we're selling an entire experience. The content we create and sell is just one segment of a thousand-piece jigsaw puzzle.

As we move to the next iteration of storytelling, as a great flattening is taking place between consumer and creator, the medium will no longer be the message. The medium will be pervasive. The message will be amateur, professional, and infinite. And it will all exist as a mutual collection of bytes, snacks, and meals.

Society has entered an interregnum, and what appears on the other side is not being decided by corporations and media giants. Consumers will have equal sway in the discussion. We need to harness this learning and help explore the future together. And as the opportunities arise to pick ourselves up and dust ourselves off—as they will—we need to understand how to evolve, how to communicate and tell stories again.

As distribution channels become extinct and irrelevant and the ubiquity of new devices gives way to truly amalgamated communications, the new commodities will be length, aggregation, immediacy, and niche.

It's not enough to sit idly by, ignoring and quieting the employee inside your company who doesn't buy CDs anymore, or canceled her cable television, or started playing video games instead of reading a book, or stopped buying the print edition of the newspaper. These people are trying to tell you about the future and how it works. It's up to you to listen.

It's time to reorganize, rethink, and get back to the business of storytelling.

Sincerely,
Nick Bilton

acknowledgments

First, I'd like to thank you, the reader, for taking the time to purchase and read this book. I hope it was informative, fun, and engaging. (If you stole this book, please think of the children, consider buying a copy, and see Chapter 6 on Me Economics.) Although hundreds of people have been involved, in one way or another, in the research, support, and creation of this book, the following are people to whom I wanted to offer an extra shout-out. (The names are intentionally jumbled; I love and appreciate everyone equally.)

Special, Special, Special Thanks

Danielle Bilton, for your patience, understanding, love, and baked goods.

Special, Special Thanks

This book would not have happened without the invaluable input from following people: David Carr, John Mahaney, Karen Blumenthal, Matthew Fishbane, Mark Hansen, Katinka Matson, John Brockman, Clay Shirky, Clive Thompson,

Larry Ingrassia, Tom Bodkin, Mike Young, John Markoff, Tim O'Reilly, Sam Sifton, Hubert McCabe, Mark Bittman.

New York Times

Arthur Sulzberger Jr., Janet Robinson, Martin Nisenholtz, Bill Keller, John Geddes, Jill Abramson, Rick Berke, Damon Darlin, David Gallagher, Suzanne Spector, Michael Zimbalist, Ted Roden, Alexis Lloyd, Justin Ouellette, Patricia McSweeney, Amy Hyde, Susan Edgerly, Brian Stelter, Jenna Wortham, Jim Roberts, Doug Latino, Kelly Doe, Brad Stone, Ashlee Vance, Steve Lohr, Matt Richtel, Miguel Helft, Tim O'Brien, Claire Cain Miller, Michael Golden, Evan "Scoop" Sandhaus, Bill Cunningham, Glenn Kramon, Rob Larson, Rob Samuels, Kevin McKenna, and Fiona Spruill.

Friends, Family, "The Book," & The Internet

The entire team at Random House, including Tina, Meredith, Jacob, Tara, Rachelle, and Jo. Emily Nussbaum, Jack Dorsey, Andrew Hearst, Joel Johnson, Dennis Crowley, Alex Rainert, Karen Bonna Rainert, Eric Beug, Dick Lipton, Naveen Selvadurai, Richard Nash, Brian Lam, Lux Alptraum, Nick Denton, Jonah Lehrer, Dan O'Sullivan, Nick Carr, Nicholas Felton, Kati London, Nora Abousteit, Bre Pettis, Tim Hannay, Steven Pinker, Dave Morin, Clifford Nass, Maria Popova, Red Burns, Tom Igoe, Anil Dash, Fred Wilson, Chloe Sladden, Max Whitney, New York University's ITP students and alumni, Linda Stone, Gideon Lichfield, Andrew Ross Sorkin, Jack Shafer, Michael Caruso, Baratunde Thurston, Frank Rose, Joe Wikert, Jimmy DiResta, Dan Gillmor, Sarah Slobin, Marshall Kirkpatrick, Chris Anderson, Mathias Crawford, Noah Robischon, the ladies and gentlemen of the Academy, Paul Berger, Kevin Slavin, Deborah Auer, Lane Becker, Jennifer Rodriguez Thor Muller, Denise & Michael, Aida & Jorge, Nancy & Sylvia, Cathy, Monica & Franky, Lissa and Debbie, Katie Cotton, Deborah Estrin, Diane Sawyer, Gillian Reagan, Nate Tabor, Zach Klein, Gary Vaynerchuk, Alicia Gibb, Andrew Savikas, Rachel D. Abrams, all the nice robots in the world, Sara Winge, Dale Dougherty, Jennifer 8. Lee, Gina Blaber, Brady Forrest, Kenyatta Cheese, Matt Buchanan, Andrea Sheehan, Scott Beale, Ori, Mor Naaman, Kim Naci, Mike Sharon, Jason Brush, Derek Gottfrid & Nick Thuesen, Jeff Koyen, Peter Ng, Bruce Headlam, Rex Sorgatz, Chad and Summer, Jennifer Magnolfi, Kio Stark, Nick Kristoff, John & Deirdre, Bob and Jamie, Ryan B., Marc and Tiff, Max and Roisin, Andrei K., Kevin E., Morgan, Leanne Citrone, Michael Citrone, Wuca & Pillow, Terry Bilton, Sandra and David Reston, Eboo Bilton and Weter, Betty and Len Bilton, Stephen, Amanda, Ben and Posh Jacobs, Daniel Jacobs, Ivan & Elsa Marin, Nathalie Marin, Chris Marin, Andy, Carm, George Jr., George Sr., Sonia, Joe, Chela, Tony, Jim, Andrea, Stephanie, Jessica, Lindsay, Diego and Yvonne, Cesar and Beatriz Southside, Sam H., Ariel Kaminer, Vint Cerf, Larry and Sergey, Tim Berners-Lee, Steve Jobs, and Bill Gates.

Smallest, But Not Least

Pixel, Hip Hop, & Magnolia.

Kthxbye!

notes and sources

The following sources represent a portion of the research and interviews used for this book. Additional links, reference papers, and interview quotes can be found online at nickbilton.com.

Introduction: cancel my subscription

5   *Canceling my subscription*: Ryan Singel, "*Times* Techie Envisions the Future of News," *Wired,* March 2009, http://www.wired.com/epicenter/2009/03/the-future-of-n. Also: Richard MacManus, "Sensors, Smart Content, and the Future of News," *Read Write Web,* March 2009, http://www.readwriteweb.com/archives/sensors_smart_content_and_the_future_of_news.php

7   *Print advertising*: Newspaper Association of America, U.S. advertising sales report.

13   *The 10-megabyte hard drive*: From an (1984) IBM print advertisement.

Chapter 1: bunnies, markets, and the bottom line

The source for some material in this chapter comes from confidential interviews with a senior-level *Playboy* manager and confidential interviews with sources close to the company; a personal interview with Jo Mason; a personal interview with Gram Ponante, a journalist covering the porn industry; personal interviews with Ollie Joone, Farley Cahen, Adella Curry of Digital Playground; and an interview on piracy with an adult entertainment industry employee.

22   *Internet and censorship*: Peter Johnson, "Pornography Drives Technology: Why Not to Censor the Internet," *Federal Communications Law Journal* 49 (1996): 217–26. Though not cited, further support comes from Jonathan Coopersmith, "Pornography, Technology and Progress," *ICON* 4 (1998).

24   *VHS won the tape wars*: Multiple news articles, including "The Beta-VHS Battle Offers Some Insights Into Coming DVD War," *The Wall Street Journal* (2006); "Porn Industry May Be Decider in Blu-ray, HD-DVD Battle," *PC World* (2006); "June 4, 1977: VHS Comes to America," *Wired* (2010); and "Porn Business Driving DVD Technology," Reuters (2005).

28   *Figures collected by AVN Media Network:* AVN is an adult industry media group.

32   *How consumers decide which adult sites they are willing to pay for:* Benjamin Edelman, "Red Light States: Who Buys Online Adult Entertainment?" *Journal of Economic Perspectives* 23 no. 1 (2009).

36   *Gawker Media*: Personal interview with Nick Denton, chief executive and founder of Gawker Media. Further interviews with Brian Lam, managing editor at Gawker Media and editor of Gizmodo.com, and Lux Alptraum, editor of Fleshbot.com and boinkology.com.

Chapter 2: scribbling monks and comic books

46   *The telephone*: "The Telephone," *New York Times*, March 22, 1876.

47   *The phonograph*: "The Phonograph," *New York Times*, November 7, 1877.

48   *Historians note that the railway brought an incredible amount of anxiety*: Personal interview with Anne Harrington, Chair and Professor for the History of Science, Harvard College. Also "The 'Railway Spine'—A New Disease," *New York Times*, October 15, 1866; Ralph Harrington, "The Railway Accident: Trains, Trauma and Technological Crisis in Nineteenth-Century Britain" (1999) http://www.york.ac.uk/inst/irs/irshome/papers/rlyacc.htm; and Ralph Harrington, "The Neuroses of the Railway," *History Today*, July 1994.

51   *One of the largest libraries in Europe*: Online library database history, Northern England.

51   *Printing press*: Elizabeth Eisenstein, *The Printing Press as an Agent of Change*, Cambridge, UK: Cambridge University Press, 1979. Also: *The Society of Printers for the Study and Advancement of the Art of Printing*, Harvard College Books Library, Boston, Mass.: 1906.

52   *Smaller, more portable books*: David Finkelstein, and Alistair McCleery, *Introduction to Book History*, London: Routledge/Taylor & Francis Ltd, 2007.

54   *Early newspaper articles described the television*: David Hajdu, *The Ten-cent Plague: The Great Comic-book Scare and How It Changed America*, New York: Farrar, Straus and Giroux, 2008.

59   *In a classic article in* Newsweek: Ken Olsen reference, *Financial World* (1976); Clifford Stoll, "The Internet? Bah!," *Newsweek*, February 27, 1995.

61   *Yet studies show that older technologies . . . emit stronger electronic waves than WiFi hubs*: Series of online articles including: Cyrus Farviar, "UK Doctor Puts the Smackdown on Wifi Fearmongers," *Engadget*, December 12, 2006; Richi

Jennings, "Wi-Fi Causes Child Cancer?," *ComputerWorld;* Collection of external links http://blogs.computerworld.com/node/5543.

61   *A wave of books*: Sven Birkerts, *The Gutenberg Elegies: The Fate of Reading in an Electronic Age,* New York: Faber and Faber, 2006; Maggie Jackson, *Distracted: The Erosion of Attention and the Coming Dark Age,* Amherst, N.Y.: Prometheus, 2009; Lee Siegel, *Against the Machine: Being Human in the Age of the Electronic Mob,* New York: Spiegel & Grau, 2008; Colleen Cordes and Edward Miller, eds. "Fool's Gold: A Critical Look at Computers in Childhood," *Alliance for Childhood,* http://drupal6.allianceforchildhood.org/fools_gold May 28, 2010.

65   *Reader's Digest:* James Playsted Wood, *Of Lasting Interest: The Story of the Reader's Digest,* Garden City, N.Y.: Doubleday, 1958.

67   *The* New Yorker *published a five-part investigative series*: John Bainbridge, "Little Magazine," *The New Yorker,* November 17, 1945: 33–42; November 24: 36–47; December 1: 40–51; December 8: 38–53; and December 15: 38–59.

68   *E. B. White captured this classic human response*: E. B. White, "Irtnog," *The New Yorker,* November 20, 1935: 17–18.

70   *Stone calls this "continuous partial attention"*: Several blog posts by Linda Stone in reference to attention and e-mail on lindastone.net.

73   *Crystal, a linguist:* "David Crystal," http://www.davidcrystal.com/David_Crystal/biography.htm.

73   *editor at large Jesse Sheidlower*: In-person interview, 2009.

75   *Research shows that they understand how to converse with different audiences*: David Crystal, *Txtng: The Gr8 Db8,* Oxford University Press, 2008; Robert Provine, Robert Spencer, and

Darcy Mandell, "Emotional Expression Online," *Journal of Language and Social Psychology,* October 2009; Interviews with Jesse Sheidlower, editor at large, North America, *Oxford English Dictionary,* 2009 and 2010.

Chapter 3: your cognitive road map

78 *Foursquare*: Dennis Crowley personal interview, March, 2010.

82 *Twitter references*: Personal interview with Jack Dorsey, co-founder of Twitter, for the *New York Times,* 2010.

84 *Imagined communities*: Benedict Anderson, *Imagined Communities: Reflections on the Origin and Spread of Nationalism,* London: Verso, 2006.

89 *Regina Lewis, AOL's consumer adviser, said:* Linnie Rawlinson, Linnie and Nick Hunt, "Jackson Dies, Almost Takes Internet with Him," CNN.com, June 26, 2009, http://www.cnn.com/2009/TECH/06/26/michael.jackson.internet/index.html

90 *A Twitter tussle*: George Packer, "Stop the World," Weblog post, Newyorker.com, January 29, 2010, http://www.newyorker.com/online/blogs/georgepacker/2010/01/stop-the-world.html. Also David Carr, "Why Twitter Will Endure," *New York Times,* January 1, 2010, http://www.nytimes.com/2010/01/03/weekinreview/03carr.html. Also: Personal blog posts on http://bits.blogs.nytimes.com.

100 *The Internet is not only breaking down barriers*: Matthew Gentzkow and Jesse M. Shapiro, "What Drives Media Slant? Evidence From U.S. Daily Newspapers," *Econometrica* 78 no. 1 (2010): 35–71; C. R. Sunstein, "The Daily We, Is the Internet Really a Blessing for Democracy?," *Boston Review* 26 (2001): 4–9.

Chapter 4: suggestions and swarms

106 *Difficulty in making predictions:* Clive Thompson, "If You Liked This, You're Sure to Love That," *New York Times Magazine,* November 23, 2008; Also: Eric Schmidt, online video from conference interview, 2010.

110 *More than half of society generally trusts complete strangers*: Rick Wilson, phone interview, 2010.

115 *The cold-start problem*: Timothy Bickmore and Justine Cassell, "Relational Agents: A Model and Implementation of Building User Trust," *CHI* 2001 3 no. 1 (2001): 396–403.

117 *"Computers as virtually infallible"*: BJ Fogg and Hsiang Tseng, "The Elements of Computer Credibility," *CHI* 99 (1999): 80–87. Also: Phone interview with BJ Fogg, Stanford University.

117 *Why people feel comfortable with well-designed sites*: "Jakob Nielsen," in-person discussion based on *New York Times* interview, March, 2010.

121 *"Swarm intelligence"*: Ashley J.W Ward, David J.T. Sumpter, Iain D. Couzin, et al., "Quorum Decision-making Facilitates Information Transfer in Fish Shoals," *Proceedings from the National Academy of Sciences* no.105.19 (2008): 6948–953. Also: Haewoon Kwak, Changhyun Lee, Hosung Park, et al., "What Is Twitter, a Social Network or a News Media?" *WWW* 2010 (2010); Gilad Lotan, "ReTweet Revolution," *ReTweet Revolution,* June 2009, http://giladlotan.org/viz/iranelection/index.html; Personal interview with Gilad Lotan, Microsoft Research Labs.

130 *Young people tended to share political news*: Brian Stelter, "Finding Political News Online, the Young Pass It On," *New York Times,* March 27, 2008, http://www.nytimes.com/2008/03/27/world/americas/27iht-27voters.11460487.html.

Chapter 5: when surgeons play video games

134 *"Is Google Making Us Stupid?"*: Nicholas Carr, "Is Google Making Us Stupid?" *The Atlantic* July–August, 2008, http://www.theatlantic.com/magazine/archive/2008/07/is-google-making-us-stupid/6868/. Also Nicholas Carr, *The Shallows: What the Internet Is Doing to Our Brains,* New York: W.W. Norton, 2010.

135 *A number of books*: Mark Bauerlein, *The Dumbest Generation: How the Digital Age Stupefies Young Americans and Jeopardizes Our Future (or, Don't Trust Anyone under 30),* New York: Jeremy P. Tarcher/Penguin, 2008. Also: Maggie Jackson, *Distracted: The Erosion of Attention and the Coming Dark Age,* Amherst, N.Y.: Prometheus, 2008.

137 *Stanislas Dehaene*: Unite de Neuroimageire Cognitive. Dehaene, Stanislas. http://www.unicog.org/main/pages.php?page=Stanislas_Dehaene.

137 *Develop a new area within the brain*: Manuel Carreiras, Mohamed L. Seghier, Silvia Baquero, et al., "An Anatomical Signature for Literacy," *Nature* 461 (November 15, 2009): 983–86.

140 *Our magnificent minds adapt*: Gary Small, Teena Moody, Prabha Siddarth, et al., "Your Brain on Google: Patterns of Cerebral Activation during Internet Searching," *American Journal of Geriatric Psychiatry* 17 no. 2 (2009): 116–26. Also personal interview with Gary Small at the SEMEL Institute for Neuroscience and Human Behavior at UCLA.

142 *Neuroplasticity*: Bogdan Draganski, Christian Gaser, Volker Busch, et al., "Changes in Grey Matter Induced by Training," *Nature* 427 (January 22, 2004): 311–32.

145 *I hear the same kinds of fears and anxieties*: "Scientists Warn of Twitter Dangers," CNN.com, http://www.cnn.com/2009/TECH/ptech/04/14/twitter.study/index.html. Also Hilary

Stout, "Antisocial Networking?" *New York Times*, July 6, 2010, http://www.nytimes.com/2010/05/02/fashion/02BEST.html and "E-mails 'Hurt IQ More than Pot'" CNN.com, April 22, 2005, http://www.cnn.com/2005/WORLD/europe/04/22/text.iq/

147 *Zettabytes:* Roger E. Bohn and James E. Short, "How Much Information? 2009 Report on American Consumers," *Global Information Industry Center*, December 2009, http://viadigitalis.org/wordpress/wp-content/uploads/2010/03/How-Much-Information.pdf. Also phone interview with researchers for the *New York Times* article and personal news article written for the *New York Times*.

149 *Surgical residents on their video game habits*: James C. Rosser Jr, Paul J. Lynch, Laurie Cuddihy, et al., "The Impact of Video Games on Training Surgeons in the 21st Century," *Archives of Surgery* 142 no. 2 (2007): 181–86.

150 *"Medical errors," which have become the eighth leading cause of death in this country:* U.S. Department of Health & Human Services, http://www.ahrq.gov. and Webmd.com.

150 *Using a Wii golf club*: Shiraz Badurdeen, Omar Abdul-Samad, Giles Story, et al., "Nintendo Wii Video-Gaming Ability Predicts Laparoscopic Skill," *Surgical Endoscopy*, January 28, 2010 and personal interviews with previous neuroscientists.

152 *Studied the newly released game Tetris:* Richard J. Haier, Benjamin V. Siegel Jr., Andrew MacLachlan, et al., "Regional Glucose Metabolic Changes After Learning a Complex Visuospatial/Motor Task: A Positron Emission Tomographic Study," *Brain Research* 570 (1992): 134–43; Richard J. Haier, Benjamin Siegel, Chuck Tang, et al., "Intelligence and Changes in Regional Cerebral Glucose Metabolic Rate Following Learning," *Intelligence* 16 (1992): 415–26; Richard J. Haier, Sherif Karama, Leonard Leyba, et al., "MRI Assessment of Cortical Thickness and Functional Activity Changes in Adolescent

Girls Following Three Months of Practice on a Visual-Spatial Task," *BMC Research Notes* 2 no. 174 (2009); and several phone interviews with Richard Haier, neuroscientist.

155 *Steven Johnson:* Steven Johnson, *Everything Bad Is Good for You: How Today's Popular Culture Is Actually Making Us Smarter,* New York: Riverhead, 2006. Also Mitchell Stephens, *The Rise of the Image the Fall of the Word,* Oxford University Press, 1998.

156 *Hand-eye reaction time*: C. Shawn Green and Daphne Bavelier, "The Cognitive Neuroscience of Video Games," in Paul Messaris and Lee Humphreys (eds.), *Digital Media: Transformations in Human Communication,* New York: Peter Lang, 2006. Also M.W.G. Dye, D. E. Baril, and D. Bavelier, "Which Aspects of Visual Attention Are Changed by Deafness? The Case of the Attentional Network Test," *Neuropsychologia*

45 (2007): 1801–811 and phone interview with Daphne Bavelier, Department of Brain and Cognitive Science and Center for Visual Science, University of Rochester, New York.

159 *Pew Research*: Amanda Lenhart, Joseph Kahne, Ellen Middaugh, et al., "Teens, Video Games, and Civics," *Pew Internet & American Life Project,* September 16, 2008, http://www .pewinternet.org/~/media/Files/Reports/2008/PIP_Teens _Games_and_Civics_Report_FINAL.pdf.pdf

Chapter 6: me in the middle

162 *Put this succinctly at a technology conference*: Kevin Slavin, Proceedings of Picnic, New York City, 2010.

171 *Movie's digital campfire*: Sitaram Asur and Bernardo A. Huberman, "Predicting the Future With Social Media," (2010), Arxiv.org, March 29, 2010, .http://arxiv.org/pdf/1003.5699.

172 *"We believe that a large portion of the people who have bought e-readers"*: Hillel Italie, "Publishers Say They're Holding

Back Some E-books," Business News, Associated Press Online, December 9, 2009.

173 Survey by L.E.K. Consulting: "Hidden Opportunities in New Media: Opportunities Uncovered and Myths Debunked," Tech., L.E.K. Consulting, January 20, 2010, http://www.lek .com/About/Hidden_Opportunities.cfm.

178 *Admitted to piracy himself*: Peter Serafinowicz, "Why I Steal Movies . . . Even Ones I'm In," *Gizmodo,* Gawker Media: May 14, 2010, http://gizmodo.com/5539417/why-i-steal-movies-even-ones-im-in

180 *Wall Street Journal pricing*: Bill Grueskin, "The case for Charging to Read WSJ.Com," *Reflections of a Newsosaur,* March 22, 2009. http://newsosaur.blogspot.com/2009/03/case-for-charging-to-read-wsjcom.html.

181 *YouTube statistics and anecdotes*: Public talk by Mike Wesch, a YouTube anthropologist, PopTech, Camden, Mass. 2009.

191 *Psychologists debated the importance of "love"*: Harry F Harlow, "The Nature of Love," *American Psychologist* 13 (1958): 673–85.

192 *Creating fake monkeys*: Harry F Harlow. and Robert R. Zimmerman, "Affectional Responses in the Infant Monkey," *Science* 130 no. 3373 (1959): 421–32.

194 *The mobile phone becomes a "transitional object"*: Rivka Ribak, "Remote Control, Umbilical Cord and Beyond: The Mobile Phone as a Transitional Object," *British Journal of Developmental Psychology* 27 (2009): 183–96.

195 *In numerous interviews, university-based human/computer interaction specialists*: BJ Fogg, phone interview, 2009. In person discussion, conference, FooCamp, Sabastapool, CA., 2009. Also phone interview with Dan Siewiorek, 2009.

Chapter 7: warning: danger zone ahead

200 *Blindness*: José Saramago, Harvest Books, 1995.

202 *While operating a vehicle*: Matt Richtel, "In Study, Texting Lifts Crash Risk by Large Margin," *New York Times*, July 27, 2009, http://www.nytimes.com/2009/07/28/technology/28texting.html.

203 *The cocktail party problem:* E. Colin Cherry, "Some Experiments on the Recognition of Speech, with One and with Two Ears," *The Journal of the Acoustical Society of America* 25 no. 5 (1953): 975–79.

206 *As research progressed in this area, key experiments found:* Broadbent is cited in Barry Arons, "A Review of the Cocktail Party Effect," *Journal of the American Voice I/O Society*, July 12, 1992.

207 *"Complete understanding . . . is still missing"*: Simon Haykin and Zhe Chen, "The Cocktail Party Problem," *Neural Computation* 17 (2005): 1875–902. Also: Interview with Kevin T. Hill, PhD candidate, Center for Mind and Brain, University of California–Davis.

208 *The attentional blink:* Jane E Raymond, Kimron L. Shapiro, and Karen M. Arnell, "Temporary Suppression of Visual Processing in an RSVP Task: An Attentional Blink?" *Journal of Experimental Psychology: Human Perception and Performance* 18 (1992): 849–60.

209 *Two very simple tasks simultaneously*: Paul E Dux, Jason Ivanoff, Christopher L. Asplund, et al., "Isolation of a Central Bottleneck of Information Processing with Time-Resolved fMRI," *Neuron* 52 (2006): 1109–120. Also: Online interview with Paul Dux, Queensland Attention & Control Lab, 2009 and phone interview with Dr. René Marois Information Processing Laboratory at Vanderbilt University, 2009.

211 *A very colorful and fun book about the brain:* John Medina, *Brain Rules,* Seattle: Pear Press, 2008. Also personal interview with John Medina, developmental molecular biologist, University of Washington School of Medicine, Department of Bioengineering, and Seattle Pacific University, 2009

212 *Multitasking pilots:* Joshua Rubinstein, David Meyer, and J. Evans, "Executive Control of Cognitive Processes in Task Switching," *Journal of Experimental Psychology* (2001).

214 *"Partial displacement theory":* Clifford Nass and Byron Reeves, *The Media Equation: How People Treat Computers, Television, and New Media Like Real People and Places,* Cambridge, U.K.: Cambridge University Press, 1996. Also personal interview Clifford Nass, Professor at Stanford University, 2009.

217 *"Multitasking Generation":* Claudia Wallis, "The Multitasking Generation," *Time,* March 19, 2006, http://www.time.com/time/magazine/article/0,9171,1174696,00.html.

217 *Study by the Kaiser Family Foundation:* "Generation M: Media in the Lives of 8–18 Year-Olds," Rep. no. 030905, Kaiser Family Foundation, March 9, 2005, http://www.kff.org/entmedia/entmedia030905pkg.cfm.

218 *Maybe you're just fooling yourself:* Eyal Ophir, Clifford Nass, and Anthony D. Wagner, "Cognitive Control in Media Multitaskers," *PNAS Early Edition* (2009), www.pnas.org/cgi/doi/10.1073/pnas.0903620106. Also phone interviews with Clifford Nass, sociologist and professor at Stanford University, 2009 and 2010.

220 *Questions related to the experiences they engage in simultaneously:* L. Mark Carrier, Nancy A. Cheever, Larry D. Rosen, et al., "Multitasking Across Generations: Multitasking Choices and Difficulty Ratings in Three Generations of Americans," *Computers in Human Behavior* 25 (2009):

483–89. Also phone interviews with Mark Carrier and Nancy Cheever, 2009.

Chapter 8: what the future will look like

229 *The* Minority Report *concepts*: Personal interview with Dale Herigstad, creative director, Schematic. Also e-mail interview with Mr. Herigstad and video by John Underkoffler about the future of user interface for 2010 TED Talk, http://www .ted.com/talks/john_underkoffler_drive_3d_data_with_a _gesture.html. Also: Wikipedia entry for *Minority Report,* en.Wikipedia.org.

234 *Test their viewing experiences on different kinds of screens*: Maria Elizabeth Grabe, Matthew Lombard, Robert D. Reich, et al., "The Role of Screen Size in Viewer Experiences of Media Content," *Visual Communication Quarterly* 6 no. 2 (1999): 4–9.

236 *Mobile phones . . . used for teaching*: Nipan Maniar, Emily Bennett, Steve Hand, et al., "The Effect of Mobile Phone Screen Size on Video Based Learning," *Journal of Software* 3 no. 4 (2008): 51–61. Also e-mail interview, December 2009.

237 *4.6 billion active mobile phones*: CTIA-The Wireless Association.

244 *Kindle:* Josh Quittner, "Will Amazon's Kindle Rescue Newspapers?" *Time,* May 5, 2009, http://www.time.com/time/ business/article/0,8599,1895737,00.html.

249 *Walter Lippmann and John Dewey*: The debate played out largely in the pages of *The New Republic,* in a series of articles dating from 1922 to 1927. Also: In-person interview, Jay Rosen, NYU School of Journalism, 2009 and in-person interview with Mitchel Stephens, author of *A History of News* and *The Rise of the Image, the Fall of the Word,* NYU School of Journalism, 2009.

255 *Cyborgs*: Gordon Bell and Steve Mann: Clive Thompson, "A Head for Detail," *Fast Company* 110, November 1, 2006, http://www.fastcompany.com/magazine/110/head-for-detail .html. Also: Personal discussion with Gordon Bell, Toronto, 2008, and personal discussion with Steve Mann, Toronto, 2008.

index

about the author

NICK BILTON is the lead technology writer for the *New York Times* Bits blog and a reporter for the paper. He writes for the *Times* about the effects technology is having on our culture and society and the sweeping changes taking place to traditional businesses. His work weaves together many different fields of storytelling, including journalism, design, technology, user interface, documentary film, advertising, and hardware hacking and how these fields will shape the future. At the *Times* he has also worked in the research and development labs, peering ten years into the future of media and helping chart the path for the future of news. Bilton is also an adjunct professor at New York University's interactive telecommunication program and speaks regularly around the world at major technology and publishing conferences and universities. He hopes to own a robot one day.